T0123116

The Train and the Telegraph

Hagley Library Studies in Business, Technology, and Politics

Richard R. John, *Series Editor*

The Train and the Telegraph

A Revisionist History

Benjamin Sidney Michael Schwantes

Johns Hopkins University Press
Baltimore

Printed in the United States of America on acid-free paper
9 8 7 6 5 4 3 2 1

Johns Hopkins University Press
2715 North Charles Street
Baltimore, Maryland 21218-4363
www.press.jhu.edu

Library of Congress Cataloging-in-Publication Data

Names: Schwantes, Benjamin Sidney Michael, 1978– author.
Title: The train and the telegraph : a revisionist history / Benjamin Sidney Michael Schwantes.
Description: Baltimore : Johns Hopkins University Press, 2019 | Series: Hagley Library studies in business, technology, and politics | Includes bibliographical references and index.
Identifiers: LCCN 2018046654 | ISBN 9781421429748 (hardcover : alk. paper) | ISBN 9781421429755 (electronic) | ISBN 1421429748 (hardcover : alk. paper) | ISBN 1421429756 (electronic)
Subjects: LCSH: Railroad trains—Dispatching—United States—History—19th century. | Telegraph—United States—History—19th century.
Classification: LCC TF563 .S39 2019 | DDC 384.10973/09034—dc23
LC record available at https://lccn.loc.gov/2018046654

A catalog record for this book is available from the British Library.

Special discounts are available for bulk purchases of this book. For more information, please contact Special Sales at 410-516-6936 or specialsales@press.jhu.edu.

Johns Hopkins University Press uses environmentally friendly book materials, including recycled text paper that is composed of at least 30 percent post-consumer waste, whenever possible.

To my parents, Carlos and Mary Schwantes

CONTENTS

Preface *ix*
Acknowledgments *xvii*

Introduction 1

1 Rights-of-Way 4

2 Dangerous Expedient 31

3 At War with Time and Space 55

4 The American System 81

5 The Struggle for Standards 103

6 Telegraphers and Regulators 120

Conclusion 147

Notes *151*
Bibliography *181*
Index *189*

The train and the telegraph in nineteenth-century America have been the subjects of countless historical studies. However, historians have either examined these significant industries in isolation from each other or have assumed that the railroad and telegraph industries enjoyed a symbiotic and trouble-free relationship during the nineteenth century. This study will demonstrate that both historical perspectives are problematic and fail to capture the complex and contingent nature of the relationship that developed between railroad officials and telegraph promoters over the course of the century.

Overview of the Book

The first chapter contextualizes the development of the American railroad and telegraph industries by comparing it with the emergence of the two industries in Great Britain a decade earlier. In comparison to American railroads, British railroad companies were well capitalized during this era because they had ready access to a large pool of wealthy investors and a liberal credit market. When British electrical telegraph promoters William F. Cooke and Charles Wheatstone sought sources of capital to finance the development of experimental telegraphs in the late 1830s, they turned to railroad officials for aid. Consequently, close personal and financial relationships emerged between British railroad officials and telegraph promoters. Telegraph promoters lauded the advantages of their communication devices for safely and efficiently managing train operations across Britain's increasingly busy rail network. In turn, British railroad officials acknowledged the validity of these arguments and invited telegraph promoters to demonstrate their new inventions. Telegraph promoters used the results of these tests to modify their experimental networks to meet the practical daily needs of railroad managers and installed telegraph lines along several British railroad rights-of-way in the early 1840s.

During the same time period, American telegraph promoters and railroad officials faced a state of affairs very different from that experienced by their British counterparts. American railroads were poorly capitalized compared to British firms and did not offer American telegraph promoters a practical source of research and development capital. They also carried a significantly lower volume of freight and passenger traffic in the 1830s and 1840s. American railroad officials managed this traffic relatively safely and efficiently through strict timetables and detailed operating protocols. They saw little need for expensive and complex electrical telegraphs that would be more likely to disrupt rail operations than facilitate them. Furthermore, American telegraph pioneer Samuel F. B. Morse's vision for his experimental device stood at odds with that of his British peers. Morse sought congressional support to build a publicly administered network that would complement the US Post Office's physical mail delivery network. The commercial potential of his invention held less interest for him, and he did not wish to place control of his patent in the hands of profit-seeking businessmen.[1] When his vision for a publicly administered telegraph network failed to materialize in the mid-1840s because of opposition from Jacksonian Democrats in Congress, Morse was forced to seek commercial partners to finance the construction of additional telegraph lines. These partners, in turn, approached railroad officials about accessing their rights-of-way. However, American railroad officials were extremely wary about placing a new, and potentially dangerous, electrical device and its necessary poles and wires in close proximity to their trains and tracks. A few well-publicized train accidents due to fallen trackside telegraph lines entangling and killing train crew members and passengers convinced many telegraph promoters and railroad officials that telegraph and railroad infrastructures did not belong together. Furthermore, some rail officials feared that permitting telegraph lines on their rights-of-way could violate their state-issued charters, since the rail firms might be perceived as engaging in a commercial activity not endorsed by state legislatures. Such actions could expose rail firms to legal liabilities and possibly lead to the revocation of their charters.[2]

Chapter 2 focuses on the decade leading up to the American Civil War. Despite telegraph promoters' limited success in building viable commercial telegraph networks along the East Coast in the late 1840s, promoters continued to expand telegraph facilities westward toward the Mississippi Valley. While a handful of railroad executives granted telegraph firms permission to access their rights-of-way, most railroad officials continued to believe that they could operate rail lines safely and effectively through time- and rule-based protocols. On a few trans-Appalachian rail lines, however, managers found rule-based operating practices insufficient to meet the increasingly complex operational requirements of their companies. In the early 1850s, the New York & Erie (Erie) and the Pennsylvania Railroad (PRR) both completed rail lines stretching from the Eastern Seaboard to the Ohio River Valley and the Great Lakes. These new lines required greater managerial oversight to keep trains running safely and efficiently because of their length. Individual managers could no longer personally supervise and direct rail activities along these lines. Instead, officials gradually began to adopt the telegraph as a tool for overseeing and controlling train movements.

Telegraphic train management challenged the rigid hierarchy of control developed by rail officials over the previous two decades by introducing human and technical intermediaries into the management structure. Senior officials depended on telegraph operators to transmit train orders to station agents in distant locations. These station agents, in turn, passed along the orders to train crews. If telegraph equipment failed to function properly, operators made errors while dispatching instructions, or station agents forgot to pass along train orders to locomotive conductors and engineers, accidents could occur. Telegraphy posed a further problem for Erie and PRR officials, since few understood the differences between the competing electrical telegraph devices of the era. They were forced to depend on outside consultants to select the most appropriate telegraph apparatus from a field that included Morse's patented design and several other rival patents. Doing so ran counter to traditional equipment adoption practices employed by American railroads, in which firms developed new equipment in-house or purchased licenses from trusted and well-known inventors close to the railroad industry.

Unlike the Erie, the PRR, and a handful of other firms, the majority of American railroads did not have the financial resources or compelling need to acquire expensive licenses to operate telegraphs and then construct trackside lines during the antebellum era. Railroad officials, however, did utilize commercial telegraph firms to send important business communications. Telegraph promoters in the Midwest recognized this opportunity and offered railroad officials free or discounted commercial telegraph services in exchange for the exclusive right to situate their telegraph lines on railroad property. Doing so enabled telegraph firms to block competitors from accessing the rail lines, thus forcing them to use less direct routes along public highways. After the Western Union Telegraph Company was organized in the late 1850s, its founders utilized the strategy quite effectively to build a substantial commercial telegraph network across the Midwest and to create a significant entry barrier for competing telegraph firms that helped to limit competition.[3]

The Civil War's dramatic impact on the American railroad industry is the focus of the third chapter. The war profoundly altered many railroad officials' views about how the telegraph could be incorporated into railroad management practices. Civilian railroad officials played a critical role in the conflict. They organized and ran the United States Military Railroad Corps (USMRR), a military organization created by the Northern government to maintain and operate railroad lines in territories occupied by the Union Army. The USMRR oversaw rail operations throughout the Eastern and Western Theaters of combat and was initially staffed by officials from the Pennsylvania Railroad. During the first year of the war, USMRR officials were also responsible for maintaining telegraph facilities along their rail lines, but this function was later assigned to a separate organization designated the United States Military Telegraph Corps. Throughout the war, USMRR officials clashed over the necessity of telegraphy for military railroad operations. Some officials argued that the telegraph was of little value for railroad operations compared with traditional time- and rule-based management practices. Other officials emphasized the importance of telegraphy and fought throughout the war to return control of the Military Telegraph Corps to the USMRR. By the end of the conflict, officials from both agencies began working closely together to manage railroad and telegraph resources. When these officials returned to civilian railroads and telegraph firms after the war, they brought with them a new outlook on how telegraphy could be used by the railroad industry.

The chapter also examines how the Civil War affected civilian railroads in the North and Midwest and led many firms to reevaluate their operational management practices during and after the conflict. Railroads in both regions were overwhelmed by wartime freight and passenger traffic. Officials recognized that dispatching trains by telegraph was necessary to resolve major traffic delays on their rail lines. Larger and better-capitalized railroads like the PRR and the Erie expanded their company-owned telegraph networks to accommodate the new dispatching requirements. Many smaller and poorly capitalized railroads began using existing commercial telegraph facilities along their rail lines for issuing train orders. This created difficulties for these companies, since commercial lines were often overwhelmed with government and civilian business, and train orders could not always be transmitted in a timely manner. Consequently, civilian railroad officials developed ad hoc dispatching practices that supplemented older time- and rule-based dispatching protocols with telegraphy but did not replace the traditional dispatching methods. Colloquially known as the "American system" of dispatching to distinguish it from alternative methods employed in Great Britain and its colonies, these dispatching practices persisted into the postwar era and constituted the basic operating guidelines of many railroads across the nation until late in the nineteenth century.

The fourth chapter addresses how the complex and ill-defined telegraphic operating practices that many railroads used in the postwar era proved to be a major source of tension within the industry. Busy railroad officials could no longer personally supervise individual train movements along their rail lines and issue train orders as needed. They depended on train dispatchers to make crucial operating decisions and keep trains running efficiently and safely across their rail networks. Dispatchers were often overworked and occasionally made serious errors that led to train collisions and other operating accidents. The American system of dispatching lacked any safeguards to prevent such errors. This exposed the rail industry to severe criticisms from public safety advocates in the 1870s. These advocates pointed to deadly collisions on rail lines where officials refused to adopt telegraphic dispatching practices as evidence that the American rail industry was technologically backward and in desperate need of internal reform and external regulation. Safety advocates argued that busy American railroads should adopt more sophisticated electrical signaling equipment, and smaller, lightly trafficked lines should improve their telegraphic dispatching practices to eliminate the inconsistencies present in the American system of dispatching. They argued that this would enable railroad firms to offer safer and more efficient transportation services to the American public.

Economic and technical factors stood in the way of railroad dispatching reform efforts in the 1870s. Few railroads had the necessary financial resources to invest in sophisticated electrical signaling equipment, particularly after the onset of a major economic depression in 1873. This financial crisis also prevented smaller, less-capitalized firms from building railroad-owned telegraph lines, as had many of the larger and better-capitalized American railroads by this time. These smaller firms depended on Western Union telegraph facilities to transmit train orders and corporate communications. Many had initially signed contracts with Western Union for free telegraph service in exchange for exclusive right-of-way access during the Civil War era. As these companies increasingly relied on telegraphic dispatching to keep passenger and freight trains moving safely and efficiently across their rail networks, they encountered serious problems with Western Union's service. Western Union's officials did not want to add additional capacity to the firm's existing telegraph network to provide additional free communication services to railroads. Instead, they criticized railroad managers for using their complementary telegraph access inefficiently. Western Union's resistance to spending capital to improve its railroad communication services forced railroad officials to economize by limiting telegraphic dispatching and depending on older, labor-intensive dispatching practices to keep traffic moving. Overworked and underpaid dispatchers, telegraph operators, and station agents bore the brunt of these decisions.

By the mid-1880s, railroad officials throughout the nation acknowledged that ad hoc telegraphic dispatching practices were no longer acceptable. Individual railroads functioned increasingly as parts of larger, multifirm rail networks, and locomotives operated by train crews from one company might routinely use rail lines belonging to another. This situation was ripe for disaster, since inconsistent train dispatching and operating practices confused crews and caused them to make mistakes that led to accidents. Chapter 5 focuses on efforts by railroad officials throughout the United States to implement operating standards on the nation's rail lines and promote uniform train dispatching practices in the 1880s and 1890s. The campaign was part of a broader industry effort to promote and enact standardization of rail infrastructure, equipment, and operating protocols during this era. In 1883, following the enactment of standard time zones across the country, participants in the General Time Convention (GTC), a railroad industry trade association with members throughout the United States, initiated a campaign to develop uniform operating rules and dispatching practices for member railroads. Over the next few years, the GTC developed standards for railroad operations that recognized the inherent necessity of telegraphic communications for railroad operations. The Standard Code established uniform dispatching practices and offered clear guidance about how train crews would operate trains across rail lines belonging to different railroads. While not all members agreed about specific dispatching protocols, most supported the new code and adopted it on their rail lines in the 1890s.

The final chapter discusses the increasingly crucial role of telegraphers in the American railroad industry and examines how concerns about the reliability of telegraphers and the impact of government-imposed regulations on hours of service led to a remarkable transformation in train dispatching practices in the first decade of the twentieth century. By the late 1890s, most railroad officials had embraced telegraphy as an operational management tool. The Standard Code offered them assurance that trains would be dispatched in a rigorous and uniform manner. However, many officials continued to be concerned about the human intermediaries within the managerial hierarchy who might inadvertently or deliberately fail to observe proper protocols. Invariably, "fallible guardians," as one industry official termed these employees, would cause serious accidents through minor mistakes or inconsistencies. In his opinion, they represented a serious flaw in the operational management practices used by American railroads.[4] Officials also faced a severe labor shortage among dispatchers and telegraphers. Skilled telegraphers were in high demand by railroad companies, but most railroads overworked them in return for mediocre wages. Few skilled operators would tolerate such treatment for long, and many gravitated toward the commercial telegraph industry after stints as railroad telegraphers.[5] As a result of these shortsighted labor policies, railroads had to constantly recruit new telegraphers to work in remote wayside stations throughout the country. The growing presence of poorly trained and often unqualified "fallible guardians" in trackside telegraph offices further justified many officials' concerns about the broader impact of telegraphy on railroad operations.

The growing shortage of qualified railroad telegraphers finally reached a critical juncture in 1907. Congress passed the Hours of Service Act to address the perceived danger posed by overworked railroad employees in critical positions. Telegraphers were limited to twelve hours of service in a twenty-four-hour period. The new rules forced railroads to effectively double their workforce of telegraphers, an impossible prospect in the early twentieth century. Major railroads responded to the act by adopting telephone train dispatching. With telephone dispatching, companies no longer needed to maintain a large pool of skilled telegraphers. Instead, they could pull railroad employees from other positions and train them quickly to send and receive standardized train dispatches by telephone.

Consequently, most major American railroads greatly reduced or eliminated telegraphic dispatching on busy rail lines by the end of the first decade of the twentieth century. While railroad officials had largely embraced telegraphy as a necessary managerial tool by the early twentieth century, and had devised protocols that enabled it to serve as an effective means for directing railroad operations, they could not control employee behavior or maintain sufficient numbers of employees to keep the network operating smoothly. Rather than attempting to resolve their labor problem, railroad officials chose to substitute a new communication device that returned power to the hands of managers. The Hours of Service Act and the railroad industry's rapid adoption of telephonic dispatching demonstrate just how tenuous the link between the railroad community and telegraphy really was, and shows how outside political, financial, and social factors shaped and mediated the relationship between the two institutions and their many managers and employees over the course of the nineteenth and early twentieth centuries.

Methodology

This volume builds on the work of historians of business management, communication, transportation, and regulation, including JoAnne Yates, Steven Usselman, and Richard R. John. In *Control through Communication: The Rise of System in American Management* (1989), Yates emphasized the critical role that innovations in communication and the dissemination of information played in the development of modern corporations. She focused on the Illinois Central Railroad (IC) in one of her chapters and drew attention to the fact that IC officials were reluctant to introduce telegraphic train management in the 1850s and early 1860s because it seemed to undermine older, hierarchical railroad management frameworks. This study expands on Yates's analysis by presenting a rigorous examination of the management practices of numerous railroads during same time period, including the Pennsylvania, the Baltimore & Ohio, the New York Central, and the New York & Erie. It shows that managerial reluctance to adopt telegraphy stemmed not only from apprehensions about how the new communication device would affect management practices but also from legitimate financial and legal concerns relating to patent and corporate law and a widespread ignorance about the fundamentals of telegraphy that lingered well into the post–Civil War era.

The book also extends Steven Usselman's political-economic analysis of technical changes and government regulation in the American railroad industry to include a more rigorous discussion of telegraphy. Usselman's *Regulating Railroad Innovation: Business, Technology, and Politics in America, 1840–1920* (2002) remains the most relevant study of how politics and regulation, internal as much as external, shaped the rail industry. The present study goes beyond Usselman's brief discussion of telegraphy in relation to signaling equipment and management practices and examines the interactions between American telegraph promoters and railroad officials in the antebellum era. It emphasizes how these interactions shaped a culture of distrust between American railroad officials and telegraph promoters that persisted into the late nineteenth century. It also details the role that commercial telegraph firms such as Western Union played in developing railroad telegraph infrastructures in the post–Civil War period and how these firms constrained internal railroad innovation.

Finally, this work expands on Richard R. John's study of politics, regulation, and the growth of the American telegraph and telephone industries in the nineteenth century. In *Network Nation: Inventing American Telecommunications* (2010), John discusses how the commercial telegraph industry gradually consolidated into a powerful and complex corporate monopoly in the form of Western Union. Many of the same "network builders" who were responsible for the growth of Western Union were also responsible for forging connections between the railroad and telegraph industries in the 1850s and 1860s. These links proved invaluable for the telegraph giant, but they also inhibited railroad technical and organizational innovation in the post–Civil War decades. In this study, I probe the political and legal factors that shaped the relationship between Western Union and railroad officials and examine how rail officials supported upstart telegraph firms that sought to challenge Western Union's near-monopolistic control of the communication industry in the United States during the closing decades of the nineteenth century.

Ultimately, this study argues that uncertainty, mutual suspicion, and cautious experimentation—rather than systematic, linear development—dominated the relationship between telegraph promoters and railroad officials in the nineteenth century. Each institution sought to maintain rigid control over its organizational practices and technical infrastructures, while at the same time harboring profound ignorance about the other's technical and organizational makeup. Furthermore, the railroad and telegraph industries functioned within an American political and legal framework that discouraged, rather than encouraged, cooperation prior to the 1870s. These economic, political, technological, and social factors created roadblocks to joint experimentation and development that officials from both industries overcame slowly and often with great difficulty.

ACKNOWLEDGMENTS

I never expected to write a book centered on the American railroad industry, a topic that I initially associated with train enthusiasts and business historians from the 1950s. However, my interest in the history of communication, particularly the telegraph, drew me into the arcane and complex realm of the railroad as I began to pursue my doctoral dissertation. I very quickly had to educate myself about railroad management and terminology to make sense of the vast collections of records I found at my disposal at various repositories, including the Hagley Museum and Library in Delaware and the Newberry Library in Chicago. Having done so, I then had to formulate an approach to studying the interconnections between two large and complex nineteenth-century industries. I benefited from the thoughtful guidance of many historians during the years that I worked on the dissertation and later the book manuscript.

As I worked on the topic, Susan Strasser helped me keep the big picture in mind and avoid writing a narrow critique of Alfred Chandler's analysis of the railroad industry. Furthermore, she demanded that my prose be readable and engaging and spent considerable effort critiquing early drafts of the dissertation. My subsequent writing has benefited immensely from her tough but thoughtful oversight. Arwen Mohun always made me keep the "so what?" question in mind as I worked on the project and guided me as I tackled questions related to the history of technology. As a third reader for the dissertation, Farley Grubb provided useful feedback from an economic history perspective, a field to which I have warmed over the years. In particular, Richard R. John provided invaluable assistance as I began work on the project, introducing me to other young scholars studying the history of communication and pointing out sources and archival collections that I had overlooked. He continued to support my research and writing as I began the process of expanding the dissertation into a book project: reading and commenting on drafts, encouraging me to see the project through to completion, and always expressing enthusiasm for my work. I very much doubt that this book would have reached the publication stage without his mentoring.

I spent many hours in the library and the Soda House archives at the Hagley Museum and Library. I owe Christopher T. Baer, Lynn Catanese, Linda Gross, Marjorie McNinch, the late Michael H. Nash, and other former and current archivists and librarians at Hagley a huge debt of gratitude for their support and assistance as I worked my way through boxes of corporate papers from the Pennsylvania Railroad and its subsidiaries and entire runs of railroad journals from the library stacks. I would also like to thank the library and archival staff at the Newberry Library, the Library of Congress, the National Archives, and the California State Railroad Museum for their assistance at various stages of the project.

I could not have completed this book without the financial support of numerous organizations. The University of Delaware and the Hagley Museum and Library offered me a Hagley Fellowship, which brought me to the University of Delaware in the first place. I also received financial support from the much-missed Newcomen Society of the United States in the form of a Newcomen Society Dissertation Fellowship. The Hagley Museum and Library's Henry Belin du Pont Dissertation Fellowship provided me with financial support and a cozy home away from home while I finished the research for my dissertation. The University of Delaware's Dissertation Fellowship enabled me to write the remainder of the dissertation and edit the full manuscript. Following the completion of my doctoral program, I received research assistance from the University of Missouri–St. Louis, the U.S. Capitol Historical Society, and the German Historical Institute in Washington, DC, while working on various aspects of the book project.

I would also like to thank the Business History Conference (BHC), particularly Roger Horowitz and Carol Lockman, for financial support to attend BHC conferences as a graduate student. The BHC welcomed me into the organization as a young scholar and served as a venue for presenting research on my project from its conception through its completion. I received critical feedback and encouragement from BHC members and participants, including Albert Churella, William J. Hausman, David Hochfelder, Pamela Walker Laird, Mark Rose, Steven Usselman, JoAnne Yates, and many, many others.

A number of scholars shared source material with me on the telegraph for the book project, including Diane DeBlois, Robert D. Harris, David Hochfelder, and Richard R. John. Andrew Bozanic and Daniel Claro provided useful feedback on early drafts of the dissertation manuscript that was later incorporated into the book manuscript.

Lastly, my parents, Carlos and Mary Schwantes, supported this project over many years. My father, in particular, enthusiastically shared his interest in the railroad industry with me, along with books and primary-source materials from his vast personal collection. He also looked over the manuscript at various stages and offered suggestions for improvement. My wife, Elizabeth Richards, married this project when she married me. Since then, she has provided encouragement and the occasional prodding as I worked to complete revisions to the manuscript. She also ensured that I continued working on the project in the hectic months and years after the birth of our daughter, Magdalena. While all the aforementioned individuals and organizations played a role in creating this work, I, alone, accept responsibility for any errors and omissions.

The Train and the Telegraph

Introduction

See how by telegraph and steam the world is anthropologized.

Ralph Waldo Emerson

The railroad and telegraph appeared to Ralph Waldo Emerson and many other nineteenth-century observers, along with later generations of historians, as powerful, transformative forces that together shaped the economic, social, and political development of the United States from the 1830s through the beginning of the twentieth century.[1] Artist John Gast's widely disseminated 1872 illustration *American Progress*, commissioned by travel writer George Crofutt for his transcontinental railroad guides, depicted a towering female figure representing the spirit of progress stringing a telegraph line along a railroad right-of-way as a smoking locomotive followed closely on her heels. The image reflected the popular notion of railroads and telegraphs as unstoppable forces of progress sweeping across the rapidly expanding nation in unison and uniting Americans under their power.[2] However, the perception (both popular and academic) of the inherent links between these two institutions has largely obscured a far more complex and contested relationship—a relationship shaped by economic, political, and social forces that created profound divisions between entrepreneurial telegraph promoters and more cautious railroad managers that required decades to mitigate, and which, in some cases, were never entirely resolved during the nineteenth and early twentieth centuries.

The unprecedented size and economic strength of these two institutions seemed proof positive to both contemporaries and scholars that railroad and telegraph networks were largely responsible for the United States' emergence as a major economic power over the course of the century. These networks had integrated the nation's isolated "island communities" into a transnational, market-based economy.[3] More than 163,000 miles of railroad track linked every part of the nation by the 1890s, and Western Union (WU), the dominant commercial telegraph company of the era, operated nearly 679,000 miles of telegraph lines throughout the continental United States. The railroad industry as a whole controlled assets worth more than eight billion dollars, in 1890s dollars, and represented the beating financial heart of America's industrial economy. The telegraph industry controlled far fewer assets, $140 million in 1890s dollars, but it too ranked in the top tier of major corporate enterprises in the United States.[4]

Travel writer George Crofutt commissioned the allegorical painting *American Progress* from artist John Gast in 1872. Chromolithograph prints of the work appeared in Crofutt's *Trans-Continental Tourist's Guide.* Gast's painting depicted the railroad and telegraph as physical manifestations of progress and helped reassure travelers that they would not be cut off from contact with the eastern United States as they journeyed westward. Gast's original painting is now in the collection of the Gene Autry Western Heritage Museum. *Chromolithograph from the Library of Congress, Prints and Photographs Division*

Despite these financial and spatial accomplishments, and both institutions' seemingly shared goal of conquering time and space to promote physical and virtual connections across the nation, industry officials maintained a wary stance toward each other throughout the era. Historian Richard White characterized the American public as being "prepared to love trains, hate trains, and be unable to live without them."[5] The same cannot be said for railroad managers' attitude toward the telegraph. Initially, railroad officials were concerned that allowing telegraph lines along their rights-of-way would expose them to unforeseen legal liabilities and safety risks. They were equally distrustful of using telegraphic communication to guide train movements along their rail lines and deal with other time-sensitive operational matters. Furthermore, early telegraph promoters tended to overpromise and underdeliver when it came to constructing reliable telegraph networks. More than a few nineteenth-century railroad officials threatened to tear down telegraph lines along their routes because of conflicts with telegraph firms over timely access to telegraph services and legal control of the lines themselves.

Over time, expediency, more than any other particular factor, drew the two industries and their leaders together, though legal, financial, political, and labor issues complicated the working relationship and at times threatened to undermine it completely. Telegraph promoters with Western Union began to see railroad rights-of-way as tools for expanding the firm's telegraph network across the country. They developed exclusive right-of-way contracts to slow the progress of rivals and consolidate regional markets.[6] Forward-thinking railroad officials began to experiment tentatively with telegraphy to monitor train movements and manage rail operations in real time across vast distances, but telegraphic train dispatching remained controversial among the majority of railroad managers. In large part, managers continued to treat the device as an unwelcome intruder because it necessitated a growing reliance on an army of "fallible guardians," the railroad telegraphers and train dispatchers whose presence disrupted traditional managerial hierarchies and reduced managers' direct personal oversight of railroad operations. Managing rail operations with the telegraph was also fraught with peril, owing to mechanical and organizational shortcomings well into the closing decades of the nineteenth century.

As railroads continued to expand in size and traffic density in the 1880s and 1890s, railroad officials faced mounting pressure from state and federal regulators to improve their safety and operating records, particularly with respect to telegraphy. Railroad managers began to develop interfirm operating standards that emphasized intensive use of the telegraph for managing train movements, but many officials continued to worry that these new management practices would strain their telegraph networks and workforce to the breaking point. After new federal regulations in 1907—championed by railroad telegraphers and other labor brotherhoods—made railroad telegraph networks more expensive to operate, railroad officials across the nation unceremoniously replaced the telegraph with a new train dispatching device, the telephone. This dénouement reveals the contingency of the relationship between the two industries and highlights the ways in which it hinged on legal, political, economic, and social factors beyond the direct control of key actors from either institution.

Rights-of-Way

> Why, I had rather have one hand car for keeping my road in repair and handling my trains than all the telegraph lines you can build.
>
> John W. Brooks, President, Michigan Central Railroad, ca. late 1840s

American painter and inventor Samuel F. B. Morse arrived in France late in the summer of 1838. He had journeyed there to obtain a French brevet, the equivalent of a patent, for his electromechanical telegraph design, consisting of "*a single circuit* and a *recording apparatus*."[1] A few weeks earlier, British officials had denied Morse's patent application, much to his great frustration, because information on his telegraph design had already appeared in a London engineering magazine.[2] Despite the setback, Morse's stay in London gave him time to meet with rival Anglo telegraph developers Charles Wheatstone and William F. Cooke and to view their new indicator telegraph in operation.[3] The previous year, Wheatstone and Cooke had demonstrated their telegraph for the directors of the London & Birmingham (L&B) Railway, who were interested in a device that would allow station agents to rapidly communicate train locations to a central dispatcher. Although the L&B's directors ultimately decided not to license the device, the inventors continued working closely with British railway officials as they improved their telegraph design.[4]

In France, Morse was received warmly. He swiftly obtained a brevet for his telegraph design and then began lobbying the French government to license the device as a supplement to the government-operated Chappe optical telegraph network. While Morse waited impatiently for a formal introductory meeting with Louis Philippe, France's popular Orléanist king, one of the directors of the St. Germain Railroad unexpectedly approached the inventor with a proposition. The railroad officials asked Morse about the practicality of constructing an electrical device that could automatically indicate train locations on the twelve-mile-long rail line.[5] Like the directors of the L&B in the United Kingdom, the French directors of the St. Germain Railroad had come to the realization that real-time knowledge of train locations would greatly benefit railroad operations and improve both safety and efficiency.

In response to the French railroad director's request, Morse sketched out a device utilizing gears and electromagnets that could indicate the presence of a train at a station by ringing a bell in a tower and instantaneously signaling the

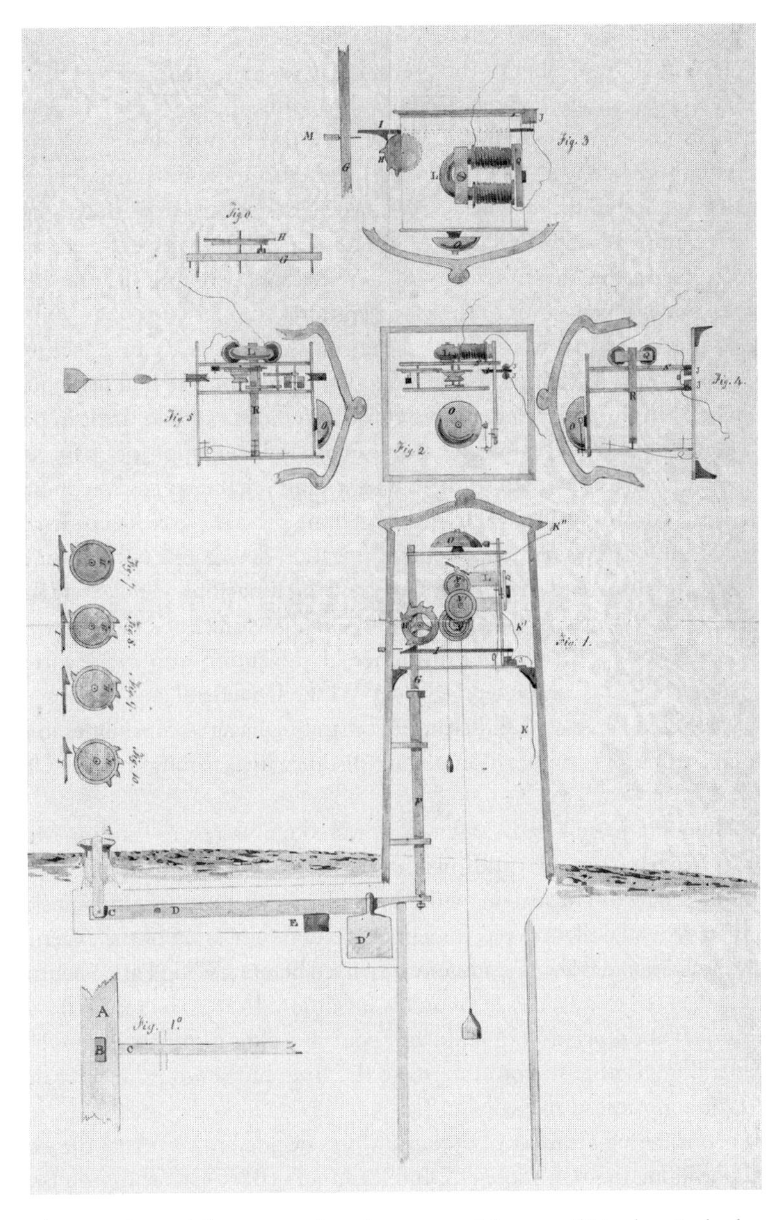

Samuel F. B. Morse did not see railroads as potential users of his electrical telegraph, but he willingly designed an electromechanical communication device for the St. Germain Railroad during a visit to France in 1838. The device would transmit train locations to each station on the line, thus allowing railroad managers to track train movements and avoid accidents on the single-track route. *"Colored sketch of railway telegraph, ca. 1838" from Samuel Finley Breese Morse Papers at the Library of Congress, 1793–1919*

train's location to every other station on the line. The St. Germain's board of directors responded favorably to the preliminary design, though a few directors balked at its cost, which Morse estimated at roughly 60,000 francs, or approximately US$11,000–12,000. Railroad officials failed to provide any funding to Morse to commence work on the project, and eventually both parties seemed to have lost interest in the scheme.[6] Nevertheless, the curiosity expressed by British and French railroad officials shows that managers in their respective countries were receptive to the idea of employing new electrical communication devices to oversee and control distant railroad operations.

Following his return to the United States from Europe, Morse focused on promoting his new invention to the federal government. He had little interest in selling or licensing his telegraph patent to American railroad firms in particular or commercial entities in general. He expressed contempt for businessmen for whom "money is the main thing to be got" and "character is of secondary & very trifling consideration."[7] He did not want private parties acquiring exclusive control of his revolutionary communication device and using it for mere pecuniary ends. Instead, he envisioned creating a nationwide electrical telegraph network, similar in concept to the French government's Chappe optical telegraph network, which would introduce his invention into every corner of the growing country. Consequently, he viewed the US federal government as the only disinterested party with sufficient capital and managerial capabilities to oversee the construction and operation of long-distance lines throughout the United States.[8]

The Democrat-controlled Congress expressed no interest in funding Morse's telegraph experiments. His rudimentary demonstrations for members of the House and Senate did little to convince his audience that the government should appropriate federal dollars to pay for a practical demonstration of the experimental communication device. Five more years passed before a Whig Party–controlled Congress decided to take a gamble on his invention. In 1843, he received a small congressional appropriation to build and operate a long-distance telegraph line. The federal government would manage the line, but it would eventually be opened for commercial message traffic.[9]

After considering a variety of options, Morse decided to construct the experimental line along the Baltimore & Ohio Railroad's (B&O) Washington Branch between the District of Columbia and downtown Baltimore. Morse chose the forty-four-mile route for burying telegraph wires because the railroad right-of-way offered a relatively straight path that was free from natural or man-made obstacles.[10] He decided to bury the wires to protect them from the elements and vandalism, perhaps in imitation of Cooke and Wheatstone, who had laid their telegraph wires underground for similar reasons. An added benefit of the railroad right-of-way was that the B&O could transport laborers and supplies for the project.[11]

The merchants and bankers who constituted the B&O's board of directors did not readily acquiesce to Morse's request for access to their property. Brothers John H. B. Latrobe, B&O company counsel, and Benjamin H. Latrobe Jr., company engineer, found Morse's project intriguing and eventually convinced the directors to consider his petition. On April 5, 1843, the board, in a split vote, authorized Morse to construct the line. The directors' authorization stipulated that the work must "be done without injury to the [rail]road and without embarrassment to the operations of the Company." Furthermore, Morse must provide the company with free access to his telegraph line and be prepared to remove the line immediately, "should it prove in any manner injurious [to the operations of the company]."[12] The arrangement revealed senior B&O officials' skeptical view of Morse's electrical communication device and their concern that it might interfere with train movements on the line or in some way prove publicly embarrassing to the rail firm and its conservative directors.

Morse's dealings with the B&O's directors illustrate a fundamental difference between the business relationships that railroad officials and telegraph promoters established in Europe and in the United States. Whereas Anglo telegraph pioneers Cooke and Wheatstone viewed Great Britain's fledgling railroad industry as potential end users of telegraphy and cooperated with railroad firms in developing their indicator telegraph device, Morse pursued an independent path. He approached the B&O for permission to use the firm's right-of-way only after he had secured congressional funding for his experimental line. Because American telegraph promoters and railroad executives shared few common business goals or frames of reference, they did not cooperate to develop the new electrical communication tool. This disconnect manifested itself in tensions between the two emerging industries and resulted in the physical separation of railroad and telegraph infrastructures throughout the eastern United States in the 1840s and early 1850s.

American railroad officials saw little use for Morse's communication device in the 1840s. By the time Morse completed his experimental line, managers in the mid-Atlantic and New England states had pioneered relatively safe and efficient operating practices for their young railroads. Influenced by US military officials, the B&O and other pioneering railroads had adopted hierarchical managerial structures and rigid time- and rule-based protocols for train movements. Consequently, as these companies began running trains along their tracks in the late 1830s and early 1840s, their managers did not believe that they needed a new and untested electrical communication device for overseeing and directing rail operations.

The Evolution of American Railroad Management Practices

In the United States, wealthy merchants and bankers from Baltimore, Boston, and other eastern port cities provided capital for early railroad projects. These wealthy individuals invested their own funds in railroad construction and served as members of the railroads' boards of directors, but they depended on skilled engineers to supervise all aspects of route surveying, railroad design, and line construction. These engineers also developed simple train operating practices for American railroads and then tailored them to fit the needs of the lightly trafficked and poorly capitalized rail lines being put into operation throughout the mid-Atlantic and New England states in the 1830s. These practices ensured safe train movements by using strict time- and rule-based management techniques from which no deviation was permitted.

In 1827, the B&O's directors submitted a request to President John Quincy Adams's secretary of war, James Barbour, for military engineers to survey the railroad's proposed route. The directors submitted their request under the auspices of the federal General Survey Act of 1824, which permitted the US government to assign military or federal civilian engineers to aid private firms engaged in internal improvement projects. Barbour assigned approximately a dozen military engineers to undertake the surveying work.[13] A few of these men, most notably then Captain William Gibbs McNeill, stayed with the B&O following the completion of the surveys and helped formulate many of the basic management practices used by the young company.[14]

On the B&O, McNeill, early in his employment with the firm, suggested financial reforms based on the practices of his former employer, the Engineer Department of the US Army. He proposed a formal, written accounting method for the B&O, including rules drawn directly from Engineer Department regulations.[15] Based on McNeill's suggestions, B&O officials directed him to prepare a "code of regulations" for the railroad in 1828. Titled *Regulations for the Engineer Department*, the primitive rule book both codified financial and administrative practices under which the railroad would be constructed and established a hierarchical operating structure.[16] As such, it provided a military-based organizational model for the young company.[17]

McNeill's *Regulations* strongly influenced early B&O management practices long after he had left the company in 1830, just as the first short section of the B&O's line opened to horse-drawn railway traffic. McNeill's rule book, however, said little about operational regulations for train movements. The earliest known guide for the drivers of the slow, horse-drawn train cars consisted of a four-paragraph broadside printed sometime in the late 1820s that offered only general principles for regulating train movements. Jonathan Knight, who succeeded McNeill as the B&O's chief engineer, continued using McNeill's practices for financial accounting and monthly administrative reporting and, presumably, the basic operating rules formulated under his leadership as well. Knight's management guide, issued under the title *Rules and Regulations*, remained in place until the end of the decade, when operational problems overtook administrative and technical issues as sources of concern for railroad officials.[18]

As railroad companies began running faster steam-powered trains along their new lines, managers quickly recognized a need for coherent and standardized operating protocols. If operational employees did not have clear and consistent guidelines to follow, railroads risked accidents whenever two or more trains occupied the same track. Managers sought to impose control over their subordinates, and coordinate sometimes complex train movements, by developing detailed employee timetables that identified meeting points for trains on single-track lines.[19] Railroad officials were pioneers in the development of printed managerial documents intended to regulate employee behavior. They recognized that strict rules and regulations offered the best means for managing employee actions.[20]

Clearly defined rules were particularly important for locomotive engineers and train conductors, since they generally operated their trains far from the observant eyes of superintendents and divisional managers. Historian Robert B. Shaw likened early trains to ships at sea; both functioned while cut off from contact with the outside world. As Shaw noted, "The basic characteristic of railroading would bring trains into violent collision in the absence of some system of control." Managers had to devise methods for controlling train movements that would prescribe specific times for each train to be at a certain location, but also provide employees with clear instructions for dealing with delays or other unforeseen problems as they operated their trains along single-track lines.[21]

The earliest efforts by officials to craft effective operational rules originated in Massachusetts. In the early 1830s, promoters began organizing a constellation of small railroads that would eventually form part of a larger rail network radiating from Boston west across the Berkshire Mountains to Albany, and south along Long Island Sound to Providence and New York City. The Boston & Worcester (B&W), Boston & Lowell (B&L), and Boston & Providence (B&P) Railroads all received charters from the state of Massachusetts in 1831 as part of this broader project.[22]

In the early 1830s, officials with the B&L and B&P recruited now Major William Gibbs McNeill and Major George W. Whistler from the B&O to serve as senior managers for both lines. McNeill and Whistler's new managerial responsibilities included devising operating protocols for both railroads. On the Boston & Providence, employees began using a set of fourteen operating rules, possibly written by McNeill, around mid-decade. When Whistler replaced McNeill as general superintendent in 1839, he developed a rule book containing forty-five detailed operating guidelines.[23]

The following rules from Whistler's 1839 rule book specified a conductor's course of action should his train deviate from the railroad's timetable. As the head of the train, the conductor exercised sole authority over the locomotive engineer and the numerous brakemen, who provided much of the grueling and dangerous labor that kept trains running safely. Brakemen coupled cars together and operated the manual hand brakes on each car while the train was in motion in order to regulate the overall speed of the train, since trains of the era lacked centralized air brakes. Whistle signals from the locomotive provided the only means for coordinating the activities of the brakemen. Most importantly, brakemen served a double duty as flagmen, a job that required the brakemen from a train to dismount and quickly run a quarter of a mile down the track in front of and behind a delayed train waving red warning flags to stop any approaching trains. Rules 15–17 from Whistler's rule book presented a clear and precise procedure that would prevent collisions between delayed trains on the B&P. The rules stipulated that once the conductor protected his delayed train by deploying the flagmen, he was required to wait thirty minutes for the closest approaching train to arrive. If the anticipated train did not arrive by then, the rules authorized him to proceed with "great care" to the nearest passing track, provided that he send brakemen ahead to scout each corner for oncoming traffic. In the event that a train broke down and could not be returned to service, the conductor was required to "send a Brakeman to the nearest town for help, with instructions to procure a horse if possible in order to proceed more quickly," as well as send out brakemen to protect his train from oncoming or overtaking traffic.[24]

As these rules illustrate, a train conductor had to take immediate action should his train become delayed or disabled. With military-like precision, he had to order his subordinates to rush out onto the right-of-way in all weather conditions to protect the front and rear ends of the train. Similarly, he had to follow a rigid course of action in the event that the train he was waiting for failed to arrive at the passing point on schedule. By waiting an additional thirty minutes to give a scheduled oncoming train enough time to reach the short, double-tracked passing point on the single-track line, he would likely avoid a head-on collision but would also significantly delay his own train's progress to the next station.

Whistler's rule book hints at the complicated nature of employee operating practices. Train conductors and locomotive engineers were expected to memorize and understand each of the forty-five rules in the rule book and were periodically tested on them. In the event of an accident, railroad managers held train crews responsible for the collision, unless they could unequivocally demonstrate that they had followed the operating protocols to the letter.[25]

As New England railroads expanded in the early 1840s, officials continued to depend on the operating practices developed by Whistler and his military cohort.[26] In 1841, the Boston & Worcester Railroad partnered with the Western Railroad Corporation, whose line connected Worcester with the eastern shore of the Hudson across from Albany, to form the Western Railroad of Massachusetts. With more than 155 miles of track, the new railroad was the longest in the United States.[27] The Western used a book of operational guidelines developed by Captain William H. Smith, George W. Whistler's brother-in-law. Smith's *Regulations for the Government of the Transportation Department of the Western Rail Road Corporation* was based largely on the operating rules used by the B&O Railroad during the 1830s. At the time, the B&O was considered by many to be the proving ground for new railroad operating methods and practices. Smith's rules had proven suitable for the Western Railroad's train operations prior to the merger, and officials saw no reason to modify them for the newly expanded line.[28]

Shortly after the merger, however, a serious accident demonstrated the shortcomings of Smith's operating protocols. On October 5, 1841, one day after the Western began running trains along its entire route, a head-on collision between two passenger trains resulted in two deaths and numerous injuries.[29] Responding to public criticism, the railroad's board of directors formed a committee to investigate the company's operating practices. With guidance from George W. Whistler, who had become chief engineer of the firm, the committee concluded that Smith's original rules were sound. Nevertheless, they offered some important revisions that they charitably characterized as "little more than amendments to the rules and regulations then in force."[30]

Crafted by Whistler, the Western Railroad's new operating rules bore a strong resemblance to the forty-five operating rules that he had developed two years earlier for the Boston & Providence Railroad. Intended to present clear operating standards, and to eliminate ambiguity wherever possible, the new rules addressed both administrative and practical operating issues. Whistler recommended reorganizing the Western's entire line into three divisions, each of which would be further divided into smaller operating units under the control of a superintendent. The head of each division would maintain a detailed record of daily operations "in order that the experience thus acquired may be rendered serviceable in subsequent operations."[31] By forcing managers to document their day-to-day operating practices and experiences, Whistler generated an important repository of data on railroad operations that was readily accessible to company officials for review.

Whistler also strongly criticized the unsystematic way that station agents communicated schedule changes to conductors and engineers. He recommended that "the time & manner of running the trains shall be established and published by the [Supervising] Engineer . . . [and] no alteration in the times of running or mode of meeting & passing of trains shall take effect, until after positive knowledge shall have been received at the office of the superintendent that [written] orders for such change have been received & are understood by all concerned."[32] This rule codified the manner in which new schedules could be issued to train crews. Whistler's rule eliminated scheduling ambiguities by ending local station agents' freedom to establish their own temporary timetables. Instead, he imposed a hierarchical framework for train control. All orders involving train schedule changes had to be cleared, in writing, through the office of the superintendent. Likewise, any train crews affected by changes in schedules had to acknowledge the changes in writing. Only after the divisional superintendent had personally approved the change could it go into effect. Whistler's rule effectively established two guiding principles for safe railroad operation: strict observance of the official timetable and the superintendent's sole authority to change train crews' operating schedules on his division or subdivision.

Whistler's 1841 operational rules for the Western Railroad, disseminated to train crews and depot staff as a printed booklet, represented a new standard for safe railroad operations in New England and the mid-Atlantic region. They established a hierarchical, rule-based management model that ensured safe and efficient train movements by placing operational authority in the hands of senior managers and by severely proscribing the ability of lower-level employees to make critical decisions or alter established protocols. The managerial principles embodied in the Western Railroad's new guidelines demonstrated "a new trust in procedures and systems, most of them involving written documents, rather than solely in the good judgment of individuals."[33] Consequently, American railroad officials indoctrinated in a strict code of managerial conduct during the 1840s and 1850s were highly distrustful of new managerial ideas or techniques that might undermine their rigid time- and document-based operating practices and give lower-level employees greater operational authority. This was an important factor that influenced American managers' negative reaction to the new high-speed communication device that emerged in the mid-1840s, the electrical telegraph.

Telegraph and Railroad Codevelopment: The British Model

British railroad officials did not share their American counterparts' rigid distrust of experimental devices and managerial practices. Well-capitalized British railroads carried significantly greater volumes of traffic than American railroads in the 1830s. Many of these rail lines—unlike those in the United States—were double-tracked for both efficiency and safety reasons, and British railroad officials were receptive to new communication devices that would allow them to manage traffic flows more effectively and prevent accidents. When Anglo telegraph pioneers Cooke and Wheatstone first began promoting their proprietary electrical telegraph in the late 1830s, railroad officials showed considerable interest.

In 1836, Cooke published a short pamphlet directed at the British railroad industry that extolled the merits of his electrical telegraph. He argued that his invention would enable railroads to constantly monitor the location of trains on their lines and reduce the costs of keeping extra locomotives at the ready to assist trains over steep inclines. Cooke also noted that telegraphic train dispatching could complement the strict time- and document-based train operating practices employed by British railways.[34]

Cooke's writings attracted attention on both sides of the Atlantic. In mid-1836, the short-lived *New York Railroad Journal*, possibly in response to his pamphlet, championed Cooke's electrical telegraph device. The editor of the journal noted that "a single track of railroad of any length can be made as effective and as safe by means of this auxiliary as any double track can be." He further added: "The advantages to railroads of this important invention can easily be understood by those familiar with railroad management. . . . We certainly think there is ample inducement for its employment upon every railroad in the United States."[35] Despite these arguments, American railroad officials chose to focus their efforts on improving rule-based operating practices rather than investing scarce financial resources in a new, unproven, and potentially costly electrical communication device.

British railroad officials, however, responded favorably to Cooke's logic. Late in 1836, he used personal connections to arrange a meeting with the directors of the Liverpool & Manchester Railway, the first exclusively steam-powered rail line in Great Britain. The railroad's managers were seeking a device to communicate between both sides of a 2,250-yard, single-track tunnel. While the directors had already selected a pneumatic "speaking tube" device for the tunnel, they offered Cooke an opportunity to field test his indicator telegraph. Cooke's telegraph used a simple alphabetical keyboard and a display panel with electrical gauges on it. The sender slowly typed out a message on the keyboard, and the receiver watched as needles on the gauges moved to indicate each letter. The design was complex and required many transmission wires, but it functioned reliably.[36]

Early in 1837, Cooke met Charles Wheatstone, professor of experimental philosophy at King's College London, who had been experimenting with a similar indicator telegraph design. While Cooke had been demonstrating his primitive electrical telegraph to the Liverpool & Manchester's board of directors, Wheatstone had independently approached the London & Birmingham Railway with a proposal "to superintend the establishment of an electrical telegraph on the railway."[37] In May 1837, Cooke and Wheatstone formed a partnership to further their joint business ventures.[38]

During the summer of 1837, Cooke followed up on Wheatstone's initial contact with the London & Birmingham Railway and conducted a number of practical demonstrations for the company's board members and its chief engineer, Robert Stephenson. Cooke successfully sent and received a message over a nineteen-mile telegraph circuit laid along the railroad's roadbed. He later repeated the test for the railroad's directors. Stephenson found the demonstration so persuasive that he proposed extending the telegraph line from London to Birmingham, a distance of roughly 120 miles.[39] While they were impressed with the telegraph line, the L&B's board balked at the complicated design of Cooke and Wheatstone's indicator telegraph and the proposed cost for laying underground cable, about £400–500 pounds sterling per mile. In October, the railroad informed Cooke that the experimental line would not be extended after all.[40]

Despite the setback, the partners took out a joint patent on their indicator telegraph in December.[41] The publicity surrounding the experimental L&B telegraph line opened doors with other railroad companies. In early 1838, Isambard Kingdom Brunel, the chief engineer of the profitable Great Western Railway (GWR), arranged a meeting between Cooke and the railroad's board. Cooke argued that, with electrical telegraphy, "the [GWR] manager in his office at Paddington [Station] would live like a spider along the line."[42] His and Wheatstone's device would permit a railroad official in a central location to contact distant station agents through a "web" of telegraph wires to obtain current information on train locations and issue train orders when necessary. The utility of telegraphy for imposing indirect managerial control over distant operations appealed to the GWR's directors. In April, the directors approved funding for a line from London Paddington Station to West Drayton Station, thirteen miles to the west. Following the line's completion, the GWR reserved the right to acquire a license from Cooke and Wheatstone for a more extensive railroad telegraph network.[43]

Morse arrived in England shortly after Cooke and Wheatstone inked their contract with the Great Western Railway. Though the GWR telegraph line would not open for another year, Morse met Wheatstone at his King's College laboratory and witnessed a demonstration of the indicator telegraph.[44] Morse remained convinced of the superiority of his own design, which used a single wire for transmission and reception instead of a dozen or more wires, but he learned a great deal about British telegraphy. Furthermore, in his meetings with Wheatstone and travels throughout London, Morse observed firsthand the growing business relationship between the Anglo telegraph inventors and the British railroad industry.[45]

While Morse returned to the United States in 1839 to improve his telegraph design and seek federal funding for an experimental government line, Cooke and Wheatstone continued working on the GWR's trackside line. It opened in July and became the first electrical telegraph in daily service. The Englishmen followed up their success with a contract from the London & Blackwall Railway (L&BR) for a telegraph line along the four-mile-long, cable-drawn railroad in the London Docklands. In England, as in the United States, early railroads experimented with a variety of options for moving railroad cars. The L&BR used stationary steam engines that pulled cars by cables instead of using expensive and complex mobile steam-powered locomotives. After the L&BR line opened in mid-1840, the telegraph quickly proved its value as a managerial tool. Station agents used it to alert senior managers when unexpectedly large numbers of passengers arrived at the London terminal, thus avoiding delays and inconveniences due to an insufficient number of railroad carriages in service. Since the L&BR used stationary steam engines, each station agent had to signal the engine operator by telegraph before the operator could engage the cable and move the cars forward. On a number of occasions, station agents used their telegraph instruments to warn engine operators of fouled cables or car derailments, allowing them to immediately disengage the engine and prevent accidents and injuries to passengers and crew. In 1849, L&BR officials publicly praised the telegraph for ensuring safe and efficient operations during its decade in service.[46]

In 1842, Cooke published a second pamphlet that was widely disseminated on both sides of the Atlantic. *Telegraphic Railways* bluntly laid out the economic and social benefits that railroad companies would reap if they incorporated telegraphy into their operations. It promoted "the safety and efficiency of Railway communication, by means not more, but less expensive than those now adopted."[47] Cooke argued that railroad telegraphy principally benefited single-track railroads by giving them "the safety and efficiency now supposed to be exclusively within the reach of double Lines."[48] Given the fact that most railroads in the United States at the time operated single-track lines, Cooke's argument should have made financial sense to American railroad officials in the mid-Atlantic and Northeast.

Cooke also noted that the telegraph offered railroad managers the ability to monitor the "state and condition of all trains on the [rail]road, at numerous given points," and by extension, the actions of subordinates such as train crews and station agents.[49] By obtaining a "bird's-eye view" of a rail line at any given moment through telegraphy, railroad officials could effectively manage train movements and impose indirect control over subordinates in distant locations. Trains would no longer travel independently like ships at sea. Instead, they would move under the direct control of local superintendents and divisional managers.

Cooke highlighted the inflexible character of railroad operations, particularly on single-track lines. Without an effective means for monitoring railroad operations from a central location, rigid employee adherence to timetables and strict running rules would be necessary to ensure safe train movements. Train crews could not pass each other at will, or deviate from their schedules, without risking serious accidents. Cooke observed pessimistically, "If it were possible that a time-table could be adhered to with mathematical accuracy, punctuality would supply the place, so far as safety is concerned, of the 'bird's-eye view' of the line."[50] However, he drew attention to Brunel's remarks on timetables and operating rules: "Upon a Railway of large traffic, there would be so many exceptions to the [operating] rule, that the rule would become useless."[51] Traffic would quickly grind to a halt on even the best single-track line if trains were forced to wait at sidings or passing tracks for delayed trains, in observance of "thirty-minute rules" and other safety regulations. Conversely, if train crews violated the rules, accidents would be the inevitable result. Cooke thus offered the electrical telegraph as a means for alleviating some of the inflexibility inherent to single-track rail lines. "If a Telegraph can be introduced into the [railroad] system, its utility will stand upon the threefold basis of Safety, Economy, and Efficiency," he wrote.[52]

In *Telegraphic Railways*, Cooke presented a convincing argument for the natural affinity between railroads and electrical telegraphy. British railroad officials on the London & Blackwall and the Great Western Railways certainly grasped many of the managerial principles that he addressed in the pamphlet. In fact, managers with both railroads were already putting these principles into practice in the early 1840s. In America, however, railroad officials did not perceive any inherent need for telegraph lines along their rights-of-way, nor could they afford to experiment with costly and unproven telegraphic communication networks as their British colleagues did.

Independent Development: The American Model

In the United States, Morse and other early telegraph promoters largely ignored the railroad industry and made no significant effort to demonstrate telegraphy to railroad officials or explain its potential utility for rail operations. Surely some railroad managers were aware of Cooke's writings about telegraphy, due to their widespread dissemination by the American railroad press, but few had any practical understanding of electrical telegraphy and how it might be implemented on their rail lines in a manner that complemented their current managerial practices. As a result, American officials continued to depend on time- and rule-based operating practices and expressed skepticism about electrical telegraphy throughout the 1840s.

Morse's decision to promote his patented electrical telegraph design solely to the federal government in the first half of the 1840s played a significant role in structuring the early relationship between the American railroad and telegraph industries during the remainder of the decade.[53] Unlike Cooke and Wheatstone in England, Morse did not establish financial partnerships with American railroad firms as he developed the first experimental telegraph device in the United States. Thus, when Morse and his partners began constructing commercial telegraph lines in the mid-1840s, they found American railroad managers indifferent toward telegraphy and hesitant to allow telegraph lines along their rights-of-ways.

Morse's 1843 conflict with the Baltimore & Ohio Railroad's board of directors regarding the placement of his experimental telegraph line on railroad property foreshadowed later disagreements between telegraph promoters and railroad officials. The B&O's board of directors, which consisted of wealthy Baltimore merchants and politicians, were apprehensive about the new communication device promoted by the painter turned inventor. The board's eventual stipulation that the line must be installed "without injury to the [rail]road and without embarrassment to the operations of the Company" illustrated officials' concerns that Morse's device with its miles of electrified wires might be dangerous and could pose a threat to their trains and passengers.[54]

Following the B&O board's unenthusiastic approval, Morse and his partners Leonard Gale, Alfred Vail, and former Maine congressman Francis O. J. Smith quickly set to work building a telegraph line late in the summer of 1843. Technical problems plagued the project. Morse had designed an underground conduit consisting of insulated copper wires encased in lead pipes. Unfortunately, his first supplier failed to meet delivery deadlines, and a second one supplied him with defective conduits.[55]

Morse also faced financial problems. Smith was a master at lining his pockets through kickbacks from contractors. Morse was constantly challenging Smith's efforts to rig bids and drive up the price of construction. Morse had only a $30,000 appropriation for the project, and Smith's schemes frittered away scarce financial resources. Furthermore, Morse feared that Smith's unabashed corruption would sully the reputation of his grand experiment.[56]

Despite these problems, Morse's team began the entrenching work in late October near the B&O's Mount Clare Station, just west of downtown Baltimore. Smith had subcontracted this work to an acquaintance, Ezra Cornell. The struggling inventor from upstate New York devised a horse-drawn plow for laying the cable underground. Cornell swiftly entrenched the first ten miles of telegraph cable by the beginning of December. Unfortunately, design flaws in the lead pipes and other technical problems forced Morse to suspend the work.

Morse spent the winter of 1843–44 in Washington, DC, reconsidering his initial plans for an underground line. By March, he had decided to build an aboveground line suspended from poles. Cornell began supervising the erection of wooden poles roughly thirty feet high and two hundred feet apart along the B&O's right-of-way.[57] After a successful seven-mile test north of Washington, Morse ordered Cornell to complete the line. By the end of May, the forty-four-mile line connected the US Capitol Building with Mount Clare

The decade following Samuel F. B. Morse's successful test of his telegraph between Washington, DC, and Baltimore, Maryland, witnessed a rapid expansion of the nation's telegraph infrastructure as entrepreneurs founded telegraph networks running north and south along the Atlantic Coast and west toward the Mississippi Valley. The majority of these lines depicted on an 1853 map by Charles B. Barr ran along public highways rather than railroad rights-of-way because of resistance from railroad officials. *Library of Congress, Geography and Map Division*

Station. On May 24, Morse demonstrated the line to a handful of dignitaries gathered in the Capitol. In Baltimore, Vail and Cornell operated their telegraph key from the third floor of a B&O warehouse overlooking the depot. Morse transmitted the preselected text "What hath God wrought!" to them, and they returned the invocation from Baltimore. The following day, the line opened for government message traffic. Commercial use followed a few weeks later.[58]

In his official capacity as superintendent of the experimental government telegraph line, Morse lobbied the House Committee on Commerce for additional funds to expand the line north to New York City. In a December 1844 report to the committee, Morse spoke glowingly of the successes of the Baltimore-Washington line during its first eight months in operation. As he briefly reported, most significant was almost certainly the first ever use of telegraphy for railroad operational management in the United States. In the spring and summer of 1844, multiday riots had broken out in Philadelphia between nativists and Catholic immigrants. During the July riots, John M. Scott, the Whig Party mayor of Philadelphia, dispatched a written report to President John Tyler in Washington, via train, seeking federal aid to suppress the rioting. When the dispatch arrived at the B&O station in Baltimore, B&O president Louis McLane directed that the telegraph be used to send an order to Washington holding a scheduled northbound train so the southbound express could reach the capital ahead of schedule with the time-sensitive document.[59]

McLane's use of the telegraph for safely coordinating two separate train movements did not establish a precedent for the B&O, or for any other American railroad. Just two years after the experiment, B&O managers struggled to find a way to reduce the increasing number of collisions on the single-track Washington Branch. Rather than using the existing Morse telegraph line along the track to report train locations and coordinate train movements—practices promoted by Cooke in his widely disseminated 1836 and 1842 pamphlets on railroad and telegraph management—the B&O's chief engineer Samuel Jones concluded that adding an expensive second track to the branch line was the only way to solve the problem. As telegraph historian Robert L. Thompson noted, "No mention of the telegraph was made among Jones' suggestions for improvement."[60]

Despite Morse's lobbying efforts in the winter of 1844–45, Congress failed to appropriate additional funds to extend his line to New York. Nor did the congressmen show any interest in Morse's proposal that the federal government buy his patent and then set up public-private partnerships in which the government would issue regional operating licenses to commercial operators. Instead, Congress temporarily placed the Baltimore-Washington line under the control of the Post Office Department.[61] Morse and his partners had to turn to private investors for funding to continue expanding their telegraph network, precisely the situation Morse dreaded. Morse saw himself as a scientist and an artist. He had neither the patience to negotiate patent licensing contracts with telegraph promoters, nor the desire to sell his telegraph patent outright on the open market. He intended for the federal government to take control of his patent and build an integrated, nationwide communication network. Congress's failure to act disrupted his plan for a federal telegraph network and left him in unknown commercial territory.

Great publicity attended Morse's public demonstrations of the telegraph during the summer of 1844. His name had become well known because of numerous publications explaining the workings of his telegraph. Consequently, he was beset with proposals from speculators offering to purchase licenses to build independent telegraph lines or to buy Morse's stake in the telegraph patent outright.[62] Fortunately for the inventor, he received the unexpected assistance of an experienced government bureaucrat, political operative, and businessman, Amos Kendall. In early 1845, Kendall, the former postmaster general under Andrew Jackson, approached Morse and offered to manage the financial interests of his telegraph patent. Morse and his partners signed over their controlling rights in the telegraph patent to Kendall. The arrangement spared them from directly negotiating licensing fees with telegraph promoters and dealing with attendant legal matters. In return, Kendall received 10 percent of the profits on the first $100,000 he earned through sales and licensing agreements, and 50 percent of all future profits. F. O. J. Smith maintained his quarter share of the patent rights separately.[63] This fractured legal arrangement meant that Kendall and Smith would each administer rights to the Morse patent and at times compete directly against each other in the race to build telegraph lines north and south along the Atlantic Seaboard and west toward the Mississippi Valley.

Rights and Responsibilities

Throughout the antebellum period, access to physical rights-of-way proved to be a major source of conflict between telegraph firms and state governments in the North and Midwest as well as between established railroad firms and fledgling telegraph companies. Railroad officials were reluctant to grant private telegraph firms and their employees unfettered access to railroad property because of possible liability for problems caused by telegraph company staff and equipment. Similarly, corporate rights proved to be another source of uncertainty and conflict in the evolving legal relationship between the two industries in the antebellum period. During this era, railroad officials, private citizens, and public officials all expressed concerns about whether railroad firms' state-issued corporate charters allowed them to enter into special contracts and grant privileges, including right-of-way access, to third parties such as telegraph companies. Many railroad officials were sensitive about engaging in contractual dealings with third parties that could be construed as going beyond the corporate powers enumerated in their company charters. Concerns about violating this corporate doctrine, known as ultra vires, influenced railroad officials' business relations with telegraph promoters.[64] Consequently, some telegraph promoters turned to state governments for legal permission to build along public highways in eastern and midwestern states. Conflicts regarding access to private and public rights-of-way and utilization of these corridors directly shaped the growth and expansion of the antebellum telegraph industry in the North and Midwest.

After Morse agreed to let Amos Kendall manage his telegraph patent rights in the spring of 1845, Kendall initially pursued a congressional buy-out strategy in line with Morse's vision. Once it became apparent to him that this approach was a nonstarter, Kendall began organizing new companies to extend telegraph service north from Baltimore through the mid-Atlantic region and into New England and south from Washington to New Orleans.[65] He chartered the Magnetic Telegraph Company (MTC) to expand the Baltimore-Washington line to New York City and took personal charge of the MTC lines as company president.[66] He also entered into a licensing agreement with Henry O'Rielly (originally spelled O'Reilly), an enthusiastic telegraph speculator from Rochester, New York, who proposed building telegraph lines from the East Coast to the Mississippi River and the western end of the Great Lakes.

As he began planning the MTC route from Baltimore north to New York City, Kendall attempted to use existing railroad rights-of-way—just as Morse had done—rather than negotiate with the owners of private turnpikes for right-of-way access or petition state legislatures for permission to string the wires along public roads. Like Morse, he soon discovered that few eastern railroad managers had any interest in telegraphy, and most did not wish to see telegraph lines strung along their railroad tracks. Instead, managers viewed him with suspicion and worried about how the new telegraph poles and wires would affect their operations.

Kendall immediately ran into difficulties with the directors of the Camden & Amboy (C&A) Railroad of New Jersey as he began negotiating right-of-way agreements in the summer of 1845 for the MTC line between Philadelphia and New York.[67] He planned to run the line from the Merchants' Exchange in downtown Philadelphia, north to Doylestown, Pennsylvania. From there, he intended to build a direct line along the right-of-way of the Camden & Amboy Railroad from Trenton to Amboy, New Jersey. The line would end at Fort Lee, across the Hudson River from New York City. The C&A's line held significant advantages. First, it provided the shortest route for a telegraph line between Philadelphia and New York. Second, Kendall hoped to take advantage of the railroad for transporting construction supplies, including thousands of wooden poles, glass or stoneware insulators, and miles of wire. Third, he sought the protection offered by the private ownership of the railroad right-of-way as a means for avoiding casual vandalism of the company's poles and wires. Lastly, since the railroad owned the entire right-of-way, he had to negotiate with only one party, rather than a multiplicity of individual landowners.[68]

Unfortunately, Kendall discovered that the senior managers of the Camden & Amboy had little interest in dealing with telegraph promoters. In a letter to his wife, Kendall complained about the ignorance and stubbornness of the C&A leadership, particularly C&A president Robert Stockton. Despite the publicity surrounding Morse and the electromagnetic telegraph, Kendall noted, the railroad's executives had "just begun to find out something about the magnetic telegraph, simple souls!"[69] He continued: "I find I have a little more patience than I once had; else I should curse and quit these New Jersey Corporations. But it is so much our interest to get along their [rail]roads that I repress my feelings; and the more readily because I am quite sure they will come to us in the end."[70]

Ultimately, Kendall could not secure a right-of-way agreement with the C&A for a reasonable rate. James D. Reid, a chronicler of the young telegraph industry, placed part of the blame on Robert Stockton's personal animosity toward Kendall for his treatment of railroad firms during his time as postmaster general in the 1830s. Kendall had instituted an express mail service to deliver news reports via horseback from New York to New Orleans in 1836, effectively cutting out railroads who charged the Post Office Department high rates for transporting newspapers quickly along various portions of the route. Stockton resented Kendall's actions, since they reduced his firm's earnings, and a decade later, the railroad president likely saw an opportunity for payback.[71]

However, Kendall's letter to his wife provides evidence that C&A managers' lack of knowledge about telegraphy likely played an equally significant—if not greater—role in the directors' refusal to give Kendall inexpensive access to their right-of-way. As Kendall had noted of his meeting with the C&A's management, managers knew little about Morse's telegraph. They did not want outside parties placing a new and potentially hazardous electrical device alongside their tracks without further study. Indeed, months later, C&A officials contacted preeminent electrical scientist Joseph Henry of Princeton University about telegraphy and sought his expert opinion about placing telegraph lines along the railroad's property.[72] Kendall's offer of free telegraph service meant little to C&A officials, since they saw no pressing need for the new device. Kendall and other telegraph promoters quickly realized that many railroad officials simply would not allow telegraph lines to be constructed along their rights-of-way until they better understood how the lines might affect railroad operations and perceived an advantage to having convenient access to the telegraph.[73]

C&A intransigence, as well as similar obstinacy on the part of the New Jersey Railroad Company, whose tracks largely paralleled the C&A's line, forced Kendall to alter his plans for the Philadelphia–New York line.[74] He complained bitterly to Morse about the delay: "In consequence of the exorbitant demands of the New Jersey and Camden and Amboy Railroad Companies, for the right of way and their control over the Turnpikes and other improvements on the direct line, much time has been occupied in discovering some other practible [*sic*] route."[75] The C&A and New Jersey Railroads' state-sanctioned monopoly over railroad transportation between Philadelphia and New York, as well as their controlling interest in private turnpikes paralleling the rail lines, led Kendall to reroute the telegraph line along public roads. Consequently, he had to obtain permission from the state of New Jersey to erect telegraph poles along its public highways. In 1845, the New Jersey legislature passed the first piece of telegraph legislation by a state government—a statute granting the Magnetic Telegraph Company "the power to build along public roads and water and imposed a penalty of $100 for damaging the firm's poles or wires." Other states adopted similar protective measures to encourage telegraph firms to build lines along public highways in their states.[76]

Frustrated by his dealings with railroad firms, Kendall considered avoiding any further negotiations with railroad officials when he began issuing bids for the final section of the MTC's Washington–New York line. The new line would connect the New York–Philadelphia line with Morse's original line in Baltimore. To circumvent dealing with the Philadelphia, Wilmington & Baltimore Railroad (PW&B), whose tracks ran directly between these cities, Kendall considered using a roundabout route from Philadelphia west to Lancaster, Pennsylvania, along a private turnpike, and then southeast along the Susquehanna River toward Baltimore.[77]

Kendall awarded Henry O'Rielly the contract to construct the Philadelphia-Baltimore line in November of 1845, despite the fact that O'Rielly was already heavily involved in constructing another telegraph line from Philadelphia to Pittsburgh.[78] O'Rielly submitted a proposal to PW&B officials offering the company free use of the new telegraph line in return for access to the firm's right-of-way. The PW&B's board considered the offer and surprisingly agreed to O'Rielly's terms in January 1846.[79]

O'Rielly's initial dealings with railroad officials went smoothly. However, ultra vires concerns surfaced shortly after his men began building the new MTC line in the spring of 1846.[80] Stephen Paschall, a Pennsylvania landowner, wrote the railroad's president expressing great displeasure and serious concerns about the actions of telegraph company employees. In a May 20 letter, Paschall strenuously objected to the construction of the telegraph line along the company's right-of-way through his property, arguing that he had not granted the railroad permission to contract with third parties for access to the land he had sold the company. He warned PW&B directors, "I consider that you are doing that which the law of incorporation never contemplated and which may eventually lead to difficulties."[81] Despite assurances by PW&B officials that the railroad's contract with the telegraph company was a routine agreement that was "severable at the pleasure of [the railroad] company" and did not in any way interfere with his rights as a landowner, Paschall refused to be pacified. In a May 24 letter, he reminded officials that the railroad's charter from the state of Pennsylvania contained no language about granting third parties contractual access to the PW&B's right-of-way. Paschall stated bluntly, "You have now assumed the right to grant to another company (or else you have become a company for another purpose) the right to erect poles for the purpose of communicating in a different manner from that contemplated in the act of incorporation."[82] Left unstated in his letter was an implied threat that the railroad's alleged violation of the doctrine of ultra vires would have serious legal consequences for the company.

In response to Paschall's letters, Edward C. Dale, the PW&B's president, vehemently denied that the railroad had in any way violated the provisions of its charter, and defended his company's right to contract with third parties for access to the railroad's right-of-way, but suggested a possible resolution to the growing legal conflict. Dale patiently explained to Paschall that MTC officials had offered to sign a separate lease agreement with the landowner granting the telegraph company a special right-of-way along the railroad's existing right-of-way. Under the new legal arrangement, Paschall, rather than the railroad, would be the one authorizing access to the land occupied by the PW&B. This would obviate the alleged ultra vires violation. Dale's convoluted legal stratagem apparently satisfied Paschall, for he did not make any further trouble for the PW&B or the MTC.[83]

While the PW&B's ultra vires concerns were resolved through patient negotiation, the company's corporate minutes hint at other difficulties between the railroad and the MTC after the line's completion. A few months after O'Rielly's workforce finished the project, the PW&B's board of directors authorized the company's treasurer to pursue claims against the telegraph company for "various charges incurred during the construction of the Telegraph." Late in 1847, the board issued a cryptic notice to the MTC "to carry out all the terms in their agreement with the company." Finally, in 1852, the PW&B refused to allow another telegraph company to use its right-of-way. The directors justified their decision by explaining that the MTC line had "caused much trouble at times—from poles falling across the track . . . and serious inconvenience to us in [situating] track and switches."[84]

Like Kendall and the MTC, F. O. J. Smith encountered numerous problems when he constructed his own telegraph line from New York City to Boston during the mid-1840s. Smith had quashed efforts by Kendall to raise money for a New York–Boston extension of the Magnetic Telegraph line. Instead, Smith raised capital from friends for his own corporation, named the New York and Boston Magnetic Telegraph Company.[85] Initially, he found New England railroad officials cooperative as he sought right-of-way agreements. His line ran north from lower Manhattan via the Harlem Railroad, then east along the public turnpike to New Haven, Connecticut. There, Smith gained access to the New Haven–Springfield Line of the Hartford & New Haven Railroad, the Western

Railroad, and finally the Boston & Worcester Railroad, which provided a right-of-way to the Merchants' Exchange office in downtown Boston.[86] Smith's contractors completed the line in June 1846, just as Kendall's contractors finished the MTC's New York–Baltimore line.

Smith's New York–Boston line reflected the miserly attitude of its founder. In its first two years of operation, the line's cheap, brittle copper wire broke repeatedly during storms. As Reid later noted, "170 breaks were reported in a section of 30 miles" of wire during one storm alone.[87] These breaks led to serious problems for Smith and his company, since the line periodically crossed over railroad tracks. At least twice, fallen telegraph poles and lines ensnared train crews. In April 1847, a Western Railroad train struck a fallen wire near Charlton, Massachusetts. The telegraph pole, still attached to the wire, swept four brakemen off the top of the train and under the trailing cars. A local newspaper detailed the carnage: "[One] was killed, another had his back severely injured, a third had his arm cut off [by the train], and a fourth had several ribs broken."[88] The next year, an accident caused by a sagging wire along the New Haven–Springfield Line in Connecticut resulted in a passenger's death.[89]

Both railroad companies and the state legislatures of Massachusetts and Connecticut blamed Smith's telegraph company for the deaths. Following the accident on the Western Railroad, the railroad's directors immediately gave orders "to cut down all telegraph poles which lean so as to endanger the lives of brakemen on the tops of trains."[90] A few weeks after the well-publicized incident, Gustavus A. Nicolls, chief engineer of the Philadelphia & Reading Railroad (Reading) of Pennsylvania, issued instructions to a telegraph company that wished to use the Reading's right-of-way: "No [telegraph] wire is allowed to hang within 20 feet of the surface of the rails—or within 4 feet of their outside line. The wire is not allowed to cross the railroad, except where the tracks change." Whether these instructions were in direct response to the Western Railroad tragedy is unknown, but they showed that railroad officials elsewhere in the United States were concerned about the potential harm that could be caused by the telegraph poles and lines along their tracks and sought to mitigate the danger.[91]

In 1848, both Massachusetts and Connecticut repealed laws originally passed to protect Smith's telegraph firm's right-of-way from vandalism, and instituted strict new ordinances "for the public safety and the protection of the rights of property owners [i.e., railroads] along the [telegraph] route."[92] The hyperbolic Smith railed against these acts to his board of directors and asked them to authorize moving the telegraph line from railroad rights-of-way to public roads. In his 1848 annual report, he questioned whether the convenience of constructing telegraph lines along railroad tracks outweighed the larger financial and social costs: "This seemingly endless catalogue of accidents and hindrances incident to the conjunction of telegraph lines with a railroad, and especially in this [business] climate, with each under a distinctive administration, and of which an impatient public takes but little count of their [legal] strictures on telegraph lines, and the great losses consequent therefrom, illustrated to the directors *the great error* involved in the preferences hitherto given to railroads over public or country roads for the site of telegraphs."[93] Smith perceptively identified a major problem that new telegraph companies faced when dealing with established railroad firms. As long as telegraph lines along railroad rights-of-way remained independently owned and operated, railroads would have little interest in assisting telegraph companies with their development and little stake in the success or failure of the firms. With no financial or managerial investment in the telegraph lines and no clear framework for directing telegraph company actions, railroad managers would be more likely to develop an adversarial relationship with telegraph promoters and view them as interfering with railroad operations. Telegraph promoters thus faced a clear choice. They could partner financially and managerially with railroad firms, as Cooke and Wheatstone had done in England, or they could avoid railroad rights-of-way entirely. Smith chose the latter option, and other New England telegraph promoters joined him by avoiding railroad routes and focusing their construction efforts on public highways instead.

The clash between early telegraph promoters and railroad managers strongly influenced the course of mid-Atlantic and New England telegraph construction in the late 1840s. Moreover, the unreliable nature of early telegraph lines, as evidenced by Smith's New York–Boston line, shaped many American railroad officials' perceptions of the value of telegraphy for railroad management.[94] Jeptha H. Wade, one of the founders of the Western Union Telegraph Company, faced strong resistance from railroad managers in the late 1840s when he tried to build a telegraph line along the right-of-way of the Michigan Central Railroad. As he later recalled, "R.R. officials claimed they not only did not want it, but could not afford to have it along their road as it would endanger both trains and passengers, and as to it being of any benefit to them that was all nonsense."[95]

Wade further noted that J. W. Brooks, president of the Michigan Central Railroad, "met with ridicule my arguments to convince him that [the telegraph] could be made useful in R.R. business, and said he was surprised that a sensible man like me should make such a claim to one of his experience." Wade remembered that Brooks exclaimed, "Why, I had rather have one hand car for keeping my road in repair and handling my trains than all the telegraph lines you can build." Brooks eventually gave Wade consent to build a commercial telegraph line along his track, but Wade later admitted that the line functioned unreliably during the seven years that it was in service and offered little practical benefit to the Michigan Central.[96]

The unreliable, legally questionable, and occasionally dangerous nature of early telegraph lines offered American railroad managers slim evidence that the telegraph might serve as a useful management tool. In fact, the telegraph appeared far less useful for directing railroad operations than the detailed framework of time- and rule-based operating practices devised by American railroad managers in the late 1830s and early 1840s. Unlike England, where railroad officials and telegraph promoters had engaged in a dialogue from the early days of the industry, the independent development of telegraphy in the United States shaped a culture of mutual exclusivity. American telegraph promoters approached railroad managers as outsiders. Since they lacked a clear understanding of railroad management practices, and railroad managers did not understand telegraphy, neither side could establish a mutually beneficial financial and technical relationship. Even when telegraph promoters and railroad officials achieved amicable relations, conflicts with private landowners along railroad rights-of-way could provoke a new set of problems with which railroad officials had to contend. Consequently, most railroad officials concentrated their efforts on improving rule- and time-based operating practices in the late 1840s, rather than encouraging the construction of telegraph lines along their rights-of-way and experimenting with telegraphy for safely managing train movements and employee actions.

Telegraphy and Timekeeping on New England Railroads

The profound division between American telegraph promoters and railroad managers in the late 1840s is perhaps best illustrated by New England railroad managers' efforts to create a standardized time zone for the railroads operating in Boston and throughout New England. At the end of the decade, more rail lines served Boston than any other city in the United States. The city housed seven separate terminals, which provided service to ten different railroads.[97] Most of these companies operated independently from one another, but they all used managerial practices based on the operating rules developed by William H. Smith and George W. Whistler in the early 1840s.

All rail companies operated their trains under strict timetables. For example, the Western Railroad and the Boston & Maine included instructions in their employee rule books forbidding train crews from running ahead or behind their daily schedule. These rules helped prevent head-on and rear-end collisions that might result from train crews running into oncoming trains or being rear-ended by following trains on single-track lines.[98]

Besides strict timetables, train crews and station agents depended on standardized railroad time zones to operate their trains safely. Many rail lines passed through communities that observed different civil time standards. Railroad officials had to ignore these local time zones and select a single time standard for the entire rail line in order to maintain operational consistency.[99]

Railroad companies terminating in Boston utilized a variety of reference points for establishing time zones along their routes. The Boston & Maine recognized the clock at their Boston depot as the standard for the line. Managers on the Eastern Division of the Western Railroad used the clock in the upper depot at Worcester, Massachusetts, as the time standard for trains operating between Springfield and Boston. West of Springfield, Western Railroad crews observed Springfield time, which was about six minutes earlier than Boston time, owing to the difference between local solar noon at each location.[100] These companies charged their conductors with keeping the clocks in local depots synchronized with their railroad's master clock by inspecting them on their first run each day.[101]

Strict timekeeping practices helped maintain safe railroad operating conditions throughout New England. Even though passenger traffic increased significantly on many regional lines in the 1840s, no major passenger train collisions occurred on any New England line between the early 1840s and 1848.[102] In early 1848, representatives from many of the major railroads in New England organized the New England Association of Railway Superintendents, the first professional railroad association in the United States. The association enabled members to share practical and scientific knowledge on subjects relating to railroad management, and it served as an umbrella organization for coordinating managerial practices among the various members.[103]

A year after the association's founding, a member from the Worcester & Nashua Railroad suggested that the organization establish a common railroad time zone for all rail firms operating in New England. This would eliminate the confusing assortment of local time standards observed by regional railroads. After eight months, the committee tasked with investigating standard time presented its findings. Committee members proposed establishing a standard railroad time zone based on a meridian thirty miles west of Boston, making New England standard railroad time two minutes earlier than Boston's local civil time. The association as a whole approved the committee's proposal and selected William Bond & Son, a Boston clock-making firm founded by the eminent clockmaker and astronomer William C. Bond, as the official timekeeper for all station clocks in Boston.[104]

In 1848, William Bond & Son acquired daily time readings from the Harvard College Observatory through a laborious process. Each day, Bond traveled to the Harvard Observatory and synchronized his pocket watch with the observatory's master clock. Then, he returned to his shop and synchronized the store's chronometer using his pocket watch. Railroads participating in the association's standardized time plan had to send station agents to Bond's Commerce Street shop on a weekly basis to synchronize their personal pocket watches to his store chronometer. In turn, the agents used their watches to reset all conductor watches and station clocks on each line. Far from foolproof, the new timekeeping plan required constant vigilance by participating companies to keep each station's master clock and each conductor's mechanical pocket watch on time.[105]

Bond and the railroad firms of Boston depended on these manual time transfers because the telegraph lines in Boston did not connect any of the city's train stations with Bond's shop or with the Harvard Observatory. By late 1849, two telegraph firms serviced Boston, F. O. J. Smith's New York and Boston Magnetic Telegraph Company and the Merchant's Line, a rival telegraph firm. Neither company provided service to Boston's seven railroad terminals. Smith's New York and Boston line had telegraph offices in Boston, Worcester, Springfield, Hartford, and New York City, all located in commercial structures, not railroad depots.[106] Similarly, the Merchant's Line avoided railroad lines along its route except for a short stretch along the Stonington Railroad between Providence, Rhode Island, and New London, Connecticut.[107] As a result, in 1849, the railroad and telegraph infrastructures in the busiest and arguably most progressive railroad city in the country existed in complete isolation from each other, because railroad officials and telegraph promoters saw no compelling reason for colocating any of their facilities.[108]

The physical separation of Boston's telegraph and railroad networks hinted at a broader commercial separation between the two industries. Boston's telegraph promoters and its railroad managers appear to have had few, if any, business interactions with each other. Even though both enterprises served similar markets, they functioned independently. Telegraph promoters in New England failed to see railroad firms as viable users of their communication device. They built telegraph lines to commercial centers rather than railroad depots. Consequently, forward-looking New England railroad managers, who might have wished to experiment with telegraphy for maintaining accurate railroad time across their growing networks, for issuing orders to subordinates, or for determining train locations on their lines, had no means to do so at the end of the 1840s.

Conclusion

Operating regulations and practices on American railroads grew more sophisticated during the 1840s, but railroad officials did not perceive an urgent need for a means of communicating along their rail lines more rapidly than the trains themselves could travel. Instead, managers improved operational safety and efficiency incrementally by developing more detailed operating rules, by implementing standardized railroad time zones, and by requiring station agents and train conductors to use synchronized clocks and watches. None of these improvements required dramatic changes in how managers conducted their rail operations, nor did they undermine the hierarchical management practices that railroad officials had developed in the early days of the American railroad industry.

British telegraph pioneer William F. Cooke's financial and technical arguments in favor of telegraphic railroad management did not resonate with American railroad managers because they faced an operating environment very different from that of the British managers. For officials managing lightly trafficked and poorly capitalized American railroads in the 1840s, Cooke's ideas must have seemed like solutions in search of a problem. Moreover, American telegraph promoters failed to demonstrate to railroad officials that telegraph lines could be an asset, rather than a dangerous and costly liability, for railroad operations. Ultimately, the American railroad and telegraph industries developed along separate, but occasionally overlapping, trajectories during this era.

Consequently, as American railroads expanded rapidly in length and traffic density at the beginning of the 1850s, American managers, unlike their British peers, lacked any means for quickly issuing orders to subordinates in distant locations or for monitoring train locations on their lines. On American railroads, information could travel only as quickly as the trains themselves.

Dangerous Expedient

> It was then a dangerous expedient to give telegraphic orders, for the whole system of railway management was still in its infancy, and men had not yet been trained for it.
>
> Andrew Carnegie

In 1850, not a single American railroad official used telegraphy for managing rail operations. By the following year, however, managers on two new trans-Appalachian rail lines, the New York & Erie (Erie) and the Pennsylvania Railroad (PRR), began placing telegraph operators in wayside stations to report train movements to a central train dispatcher and to pass along operating instructions from superintendents to station agents and train crews. Two important factors led to the sudden implementation of telegraphic dispatching. First, telegraph promoters and railroad managers began engaging each other in a dialogue about the positive role that telegraphy could play in railroad management, and the importance of railroad rights-of-way for the expansion of telegraph lines in the East and Midwest. Second, Erie and PRR managers began seeking new ways to move trains safely and efficiently along rail lines that stretched hundreds of miles through the rugged Appalachian Mountains. Like British railroad officials in the 1840s, they began to recognize that telegraphy could provide a useful tool for gathering information on train movements and imposing order on distant railroad operations. These two factors led managers on the Erie, PRR, and eventually a number of other long-distance rail lines to begin experimenting with telegraphic train dispatching in the early 1850s.

Despite the sudden emergence of telegraphic train dispatching on a few long-distance lines at the beginning of the decade, many antebellum railroad officials managing smaller rail lines remained skeptical of the new device. No consensus emerged among these officials about the proper role of telegraphy in railroad management. Some worried that telegraphy would make rail operations more dangerous because it would introduce additional human and nonhuman intermediaries beyond their direct control.[1] Others believed that telegraphic dispatching was simply unnecessary on their lightly trafficked rail lines. Moreover, few American railroads had the financial resources necessary to secure a license for Morse's or one of his rival's telegraph patent and construct their own lines during this decade. Consequently, most railroad officials continued to depend on traditional managerial tools, such as timetables and strict operating rules, for guiding train movements. They avoided telegraphic dispatching whenever possible, or reserved it for extraordinary circumstances.

A Disruptive Influence

In 1850, Andrew Carnegie, the future steel magnate, secured a job as a messenger boy for the Pittsburgh office of the Atlantic & Ohio Telegraph Company (A&O). The company organized by Henry O'Rielly had completed a line from Philadelphia to Pittsburgh in the late 1840s, during the boom in telegraph line construction that swept the United States after Samuel F. B. Morse completed his experimental Baltimore-Washington line.[2] Carnegie began by delivering telegrams for the A&O, but he gradually improved his skills as a telegrapher and achieved the position of assistant operator for the Pittsburgh office in 1852.[3]

Carnegie's competence as an operator attracted the interest of Thomas A. Scott, the newly appointed superintendent of the Pennsylvania Railroad's Western Division.[4] The PRR had finished its main line from Philadelphia to Pittsburgh at the end of 1852, making it only the second company in the country to complete a trans-Appalachian route.[5] Scott oversaw rail operations on the mountainous, one-hundred-mile stretch of single-track line west of Altoona, Pennsylvania. During his first few months as superintendent, Scott had spent a great deal of his time in the A&O's office sending daily reports to his superiors in Altoona and Philadelphia. Frustrated by his constant trips to the telegraph office, Scott quickly determined that the PRR needed better access to telegraph facilities in Pittsburgh. Early in 1853, he appealed to J. Edgar Thomson, the PRR's president, for permission to run a branch telegraph line from the A&O's main office to the railroad's Pittsburgh terminal and to hire a personal telegrapher and clerk. Thomson approved the proposal, and Scott offered the position to Carnegie.[6]

In a letter to a relative, Carnegie described his new job: "There is not much telegraphing to do, but it is necessary for [the railroad] to have an office—The line runs alongside of the [track] & as there is only one track laid yet—the time the different trains pass stations must be known."[7] Carnegie's note reveals that PRR managers were monitoring train locations telegraphically on the line between Philadelphia and Pittsburgh in 1853.

In fact, Pennsylvania Railroad officials had been monitoring train locations telegraphically on portions of their line since 1851. The company's sixth annual report, dated January 1852, contained the company's first published reference to telegraphy. Superintendent of Transportation Herman Haupt noted that delays in service along the length of the railroad's single-track lines were to be expected when trains departed from their assigned schedules. This problem became "greatly aggravated by the fact that . . . express passenger trains pass over the [rail]road in both directions at night, when the telegraph line is not in operation, and when no information can be obtained of the relative positions of trains running in constant expectation of meeting."[8] Haupt's report illustrates the limited extent to which telegraphy was employed on the PRR in its first year of operation. As the report noted, Pennsylvania railroad managers found telegraphic communication to be a helpful, but not an essential, aspect of operational management. After A&O agents closed their trackside telegraph offices for the evening, railroad managers expected train crews to depend on the older and more fundamental tools of railroad management—time- and rule-based operating protocols—that had been developed during the previous decade.

Carnegie's later recollections about his work for Scott further illustrate PRR officials' distrust in the telegraph in the early 1850s. As Carnegie recalled: "The railway was a single line. Telegraph orders to trains often became necessary, although it was not then a regular practice to run trains by telegraph. No one but the superintendent himself was permitted to give a train order on any part of the Pennsylvania system, or indeed of any other system, I believe, at that time. It was then a dangerous expedient to give telegraphic orders, for the whole system of railway management was still in its infancy, and men had not yet been trained for it."[9] Indeed, Carnegie was instructed to issue telegraphic train orders only when Scott was present to approve each one personally, since he was superintendent of the Western Division and thus directly responsible for all railroad operations under his purview. This restriction limited the possibility that Carnegie, or another telegraph clerk, might issue conflicting orders to different train crews and thereby precipitate a collision.

Early in his career with the PRR, Carnegie knowingly tested the company's restriction on subordinates issuing train orders. Scott was late for work one morning when Carnegie arrived at the Pittsburgh depot. The young clerk discovered that an accident on the lines east of Harrisburg had sidetracked all of the freight traffic on the western part of the PRR main line. Passenger trains were proceeding forward at a crawl, since their conductors had to send brakemen forward at every sharp curve to wave warning flags in case of oncoming traffic. After evaluating the situation, Carnegie began issuing orders in Scott's name and eventually cleared all the stalled trains from the Western Division. When Scott arrived at the depot, Carnegie confessed his actions. Although he did not recall Scott admonishing him for his presumptuousness, Carnegie knew that Scott was "afraid to approve what I had done."[10]

While the passage of time may have colored Carnegie's memories of his early days with the PRR, his story contains an important truth. In the United States, even the most technologically engaged railroad officials remained deeply ambivalent about the role of telegraphy in railroad operations during this era. A few officials like Scott were willing to use telegraphy for monitoring train locations and issuing orders to conductors and engineers, but most were fearful of the disruptive consequences of the new communication device that seemed to challenge the fundamental railroad management protocols and managerial hierarchy that had been developed through trial and error in the 1830s and 1840s.

Telegraphic Standards

Railroad officials were further confused and frustrated by the proliferation of competing telegraph devices in the early 1850s. During the nineteenth century, railroad firms adapted new equipment such as improvements to locomotives, rolling stock, rails, and so on through processes both within the railroad industry and in the marketplace.[11] In certain situations, railroad managers and employees developed incremental improvements to existing equipment as circumstances demanded. These improvements went into an "inside" pool of unpatented equipment from which any railroad could borrow. When insider innovations could not meet their needs, railroad managers had to turn to the marketplace. This course of action involved a much more complicated and expensive process of negotiating licensing agreements with inventors. Managers often did not fully understand the operating features of these marketplace inventions and had to depend on consultants, or the inventors themselves, for detailed information on the new devices. In the case of telegraphy, railroad managers were entirely at the mercy of early telegraph promoters when they sought technical and financial information on competing telegraph devices.

When railroad managers on the Erie, the PRR, and other long-distance rail lines began investigating telegraphy in the early 1850s, they entered a marketplace rigidly divided by three patents. Morse's business agent, Amos Kendall, and Morse's partner, Francis O. J. Smith, controlled the rights to Morse's magnetic telegraph design, which transferred information via Morse code along a single electrical circuit. An electromagnetic register recorded the code on a continuously moving tape with a pen or pencil. Morse's device offered a means of rapid and dependable communication, but it required trained operators familiar with his eponymous code.[12]

In the late 1840s, Scotsman Alexander Bain obtained a US patent for a telegraph that used an electrochemical process to make marks on a recording tape using a code somewhat similar to Morse's. In 1854, the United States Supreme Court ruled that Bain's patent infringed on Morse's telegraph patent.[13] This decision forced all telegraph companies using the Bain patent to close down, or merge with competitors using the Morse patent. However, during the years that the Bain patent was recognized, it offered a less expensive alternative to the more costly Morse patent.

Finally, inventor Royal House of Vermont developed a third telegraph device shortly after Morse debuted his design. The more complicated House apparatus, which bore a resemblance to William F. Cooke and Charles Wheatstone's indicator telegraph, used a keyboard on which messages could be typed using a standard alphabet. On the receiving end, the text was automatically printed on a paper tape. Untrained operators could use the House device, but it was less reliable than Morse's telegraph over long distances and more prone to mechanical malfunctions.[14]

Railroad managers thus had to carefully evaluate the relative strengths and weaknesses of these three competing designs as they contemplated equipping their rail lines with telegraph facilities. The high cost involved in licensing telegraph patents, combined with the limited financial resources of most American railroads, made the selection process all the more important. Some managers became personally involved in evaluating each telegraph design, while others depended on consultants to pick the best telegraph for their railroads. As a result, the choices that managers made in the early 1850s often depended as much on the personal and professional relationships that they established with telegraph promoters as on the relative technical merits of each design.

First Movers

Fortune smiled on neither the New York & Erie Railroad nor the New York and Erie Telegraph Company during these firms' early years. Yet, each of these hard-luck companies played a significant role in advancing railroad management practices in the United States at the beginning of the 1850s. The Erie became the first railroad in the United States to dispatch trains telegraphically, owing largely to the close personal relationship between Charles Minot, the Erie's general superintendent, and Ezra Cornell, the head of the telegraph company. This relationship strongly influenced Minot's decision to take a profound risk in 1851 and incorporate telegraphy into the Erie's managerial practices.

New York City merchants received a state charter for the Erie Railroad in 1831. These merchants hoped to take traffic away from the Erie Canal by creating an alternative east-west transportation route across New York's Southern Tier. The railroad's founders faced many challenges before they completed their line from the Hudson River to the village of Dunkirk on Lake Erie. The War Department controlled by Jacksonian Democrats refused to provide government engineers to survey the route. Directors failed to raise sufficient capital to meet the requirements stipulated in the company's founding charter. Furthermore, a provision in the charter prohibited the company from building outside New York State and left its terminal—at least initially—nearly twenty miles upriver from Manhattan on the west bank of the Hudson. The same provision resulted in additional problems when state surveyors discovered that the best route for the new rail line passed through northeastern Pennsylvania. Eventually, company leaders resolved all these problems, and construction began on the railroad in 1836.

The first train ran along the lower part of the line in 1841, but ten more years elapsed before the line was completed to Dunkirk.[15]

In 1850, the Erie's board of directors appointed Minot superintendent of the railroad.[16] A Harvard-trained lawyer, Minot had briefly practiced law before taking a position as superintendent of the Boston & Maine Railroad (B&M) in the early 1840s. At the B&M, he developed a reputation as a "most capable and progressive" administrator.[17] He was one of the founding members of the New England Association of Railway Superintendents and had served on its Standard Time Committee.[18] At the time of his selection as superintendent of the Erie, the editor of the *American Railway Times* of Boston commented that Minot was "one of the most accommodating and efficient railroad officers in the country," and that the Erie's directors had "secured to their [rail]road a most valuable man."[19]

Early in his career with the Erie, Minot lived up to his reputation as a forward-looking and capable manager. As he was supervising the final phases of construction of the Erie's line across southern New York, Minot became friends with telegraph promoter Ezra Cornell.[20] Since his pioneering construction work on Morse's Baltimore-Washington telegraph line in 1843–44, Cornell had been involved with telegraph line construction throughout the East. In 1845, he received a contract from businessmen in Utica, New York, to construct a line from Manhattan to Buffalo using the Morse patent. After completing the New York, Albany & Buffalo (NYA&B) telegraph line in 1846, Cornell briefly served as its superintendent until he had a falling out with the line's owner, Theodore Faxton.[21] Cornell left the company and began organizing the Erie and Michigan Telegraph Company to build a line from upstate New York to Detroit.[22]

As a result of his difficulties with Faxton and the NYA&B, Cornell sought an alternative route for a trunk line to connect the Erie and Michigan Telegraph Company with the profitable New York City market. When Amos Kendall refused to issue him a Morse license to compete against the successful NYA&B, Cornell obtained one from F. O. J. Smith, who shared Cornell's distain for Kendall and Faxton.[23] In 1847, Cornell incorporated the new telegraph firm and quickly went to work constructing a line along public roadways paralleling the future right-of-way of the New York & Erie Railroad.[24]

Cornell's New York and Erie Telegraph Company proved to be a disappointing venture. The line was poorly constructed and managed. It functioned so erratically that managers from Cornell's own Erie and Michigan Telegraph Company at one point refused to exchange message traffic with the New York and Erie line until its service improved.[25] Industry observer James D. Reid later called the line a "perpetual failure from the start. It was a great artery, but had no faculty for propelling blood."[26] Recognizing the precarious financial state of the company, Cornell began looking for ways to salvage his investment.

As Cornell was struggling to save his failing telegraph firm, Minot was trying to convince the directors of the Erie Railroad to permit him to build a telegraph line along the railroad's single-track right-of-way. Minot had taught himself Morse code during his time as superintendent of the Boston & Maine Railroad and was familiar with the basic principles of telegraphy and the competing telegraph devices in the marketplace.[27] Possibly influenced by his growing friendship with Cornell, Minot decided that the Erie Railroad needed a telegraph line under its control for business purposes. He eventually persuaded the directors to authorize the construction of a line "without reference to patents, and without determining the [telegraph] machinery to be used" upon its completion.[28] In theory, this clause gave him freedom to negotiate a favorable licensing agreement with the Morse, Bain, or House patent holders for transmission equipment after he had already erected telegraph poles and wires. In reality, though, his plans revealed a clear preference for Morse equipment.

In mid-1850, railroad crews began erecting Minot's telegraph line. Cornell loaned Minot glass insulators and Morse telegraph equipment for the line at no cost to the railroad.[29] Cornell's actions angered Smith, who viewed the railroad's telegraph line as a direct competitor with the New York and Erie Telegraph Company line, in which he was an investor. Minot offered to purchase rights to the Morse patent from Smith, but he refused. Instead, Smith wanted the Erie Railroad to acquire stock in the failing commercial telegraph company and keep it from going bankrupt.[30] Minot politely declined the offer and wrote to Smith that the Erie "would make arrangements with the New York and Erie Telegraph Company to work [the railroad line] for us."[31]

Even before the railroad telegraph line's completion in early 1851, Minot had begun recruiting telegraph operators for the Erie's stations. In December of 1850, Minot wrote to David H. Conklin, a young operator on Cornell's commercial line, asking him for help with the railroad's line. In the letter, Minot noted that "the telegraph would be useful in operating a railroad, as it was hoped that it could be utilized not only by showing the location of trains at all times but in [facilitating] the movement of trains." He concluded by requesting Conklin's help because "the work of erecting the Erie line was dragging, which was exceedingly annoying to the writer, as he had been principal in advising its adoption."[32] Minot's statements indicate that he was aware of the potential value of the telegraph as a tool for monitoring railroad operations in real time. More importantly, the Erie's telegraph line provided him with an opportunity to impose direct managerial control over the actions of train crews and station agents across the entire Erie route.

Minot had an opportunity to put his ideas for operating trains via the telegraph into practice in the fall of 1851. The Erie Railroad had been completed between New York and Dunkirk for only a few months, and conductors were running their trains based on strict schedules and the company's detailed rule book. The Erie's rule book stipulated that trains had to wait an hour for delayed oncoming traffic at passing points on the single-track line before they could proceed with caution.[33] One day in the fall, Minot happened to be aboard a westbound express train from New York City.[34] According to the timetable, his train was scheduled to pass the eastbound express at Turner's, a station approximately fifty miles from New York. After waiting a few minutes, Minot concluded that the eastbound train was late. Rather than wait an hour before proceeding, he asked the station agent to contact the next station at Goshen, New York, and see if the express had reached that stop yet. After learning that it had not, he telegraphed the agent at Goshen to hold the oncoming express, and asked his train's conductor and engineer to proceed in violation of the company's operating rules. According to later recollections of the event, his engineer steadfastly refused, supposedly declaring, "Do you take me for a damn fool? I won't run by that thing!"[35] Ultimately, according to Erie historian Edward Mott, Minot took control of the train and ran it himself. At the next station he repeated the process, until his train eventually passed the eastbound express farther up the line. In all, Minot's actions saved over an hour in delays and demonstrated the usefulness of telegraphy for overseeing and managing train movements.[36]

Minot's pioneering use of the telegraph for issuing train orders did not sit well with the Erie's train crews. Neither engineers nor conductors liked the idea of telegraphic dispatching, as it undermined the strictly enforced operating rules and regulations that they had been trained to observe. As Mott recalled, engineers and conductors deemed telegraphic dispatching "unsafe and unwarranted by rules governing operating." The Erie's engineers "resolved that they would not act upon an order sent by wire, and would not run their trains against a ruling time-card train," as Minot had done in his test.[37] In response, Minot issued orders to the railroad's employees clearly stating that only senior managers had authority to issue train orders that ran contrary to the Erie's timetable or operating rules. This action apparently calmed the nerves of the train crews and eased their transition to the new telegraphic dispatching methods.[38]

After his 1851 dispatching experiment, Minot began organizing the first telegraph department for an American railroad. In January 1852, he finally acquired a Morse patent license after Cornell's New York and Erie Telegraph Company went bankrupt.[39] A few months later, he appointed nineteen-year-old telegrapher Luther Tillotsin as superintendent of the new telegraph department. By 1853, Minot had created the most advanced railroad telegraph network in the United States.[40]

Minot's revolutionary telegraphic train management model improved operational efficiency and safety, but it did not render the Erie, or its partner railroads, immune from collisions. Early in 1853, the Erie acquired control of the

Paterson & Hudson River Railroad to give it access to Jersey City and New York harbor. In May, a southbound Erie express and a northbound Paterson Railroad train collided head-on near the Jersey City rail yard. Subsequent investigations showed that the Erie train had been running according to a new timetable that went into effect the day of the accident. The Paterson train's conductor was still observing the old schedule. A few months later, another collision took place within a mile of the original crash site under similar circumstances. While the Paterson Railroad had installed a telegraph line along its right-of-way prior to 1853, the company's managers were not using the line for train control at the time of the collisions. Minot's innovative managerial practices had clearly not spread to the Paterson Railroad or influenced its managers' operating protocols.[41]

The New York & Erie Railroad's annual report for its first two years of operation further emphasized the limits of Minot's telegraphic innovations. In the 1853 report, the directors credited the Erie's telegraph network with giving "a feeling of security, to the managers and operatives of the road, against a large class of accidents, to which, without it, they are particularly exposed."[42] However, the directors emphasized that the telegraph saw service primarily when excess seasonal traffic caused delays and backups on the rail line. As the report noted, "The telegraph is only permitted to be used for [train dispatching], when the trains have become deranged, and then only by one person, on each division, specially authorized to perform this duty."[43] These comments highlighted the limited nature of telegraphic dispatching practices on the Erie. Except in extraordinary circumstances, timetables and strict operational rules, rather than telegraphic train monitoring and dispatching, governed routine train movements on the Erie's main line. Given the mixed messages expressed by Erie Railroad officials in their annual report, it is easy to understand why other railroad managers in the United States remained skeptical about the value of telegraphy for railroad management.

The Pennsylvania Railroad and the Telegraph

On the Pennsylvania Railroad, one of the Erie's major rivals in the early 1850s, the managerial decision to build a railroad telegraph line stemmed from extensive research and debate (perhaps not surprising, given the firm's eventual reputation as the standard setter for the railroad industry), rather than a personal relationship between railroad officials and telegraph promoters. The Pennsylvania's senior officials did not commit their limited financial resources to a specific telegraph device until they had thoroughly investigated the relative merits of each design. A committee, instead of an individual, mediated the PRR's decision to invest in railroad telegraphy.

The Pennsylvania Railroad was completed just over a year after the New York & Erie Railroad. In 1846, Philadelphia merchants grew concerned that the Baltimore & Ohio Railroad planned to make a connection with Pittsburgh and funnel the resources of western Pennsylvania through Baltimore, bypassing Philadelphia. In response, the Pennsylvania legislature chartered the Pennsylvania Central Railroad (later simply referred to as the Pennsylvania Railroad) and permitted the company to begin work on a railroad line between Harrisburg and Pittsburgh.[44] After sufficient funds had been raised to meet the terms of the company's charter, the company's directors selected J. Edgar Thomson as the chief engineer for construction of the line.[45]

After his appointment in 1847, Thomson quickly set to work surveying the proposed route and issuing construction contracts in order to meet deadlines in the firm's state-issued charter. From Harrisburg, the PRR's line wound its way westward into the Allegheny Mountains of central Pennsylvania. For financial reasons, the company had to depend on the state of Pennsylvania's preexisting portage railroad to move passengers and cargo over the peak of the Alleghenies. The portage railroad consisted of a series of stationary steam engines that pulled canal boats up inclined planes on wheeled carriages. It was slow and forced the PRR to transship all passengers and cargo at either end of the portage railroad line.[46] By late 1852, the PRR had completed the eastern and western sections of its line. The only incomplete section remained the difficult portion through the Alleghenies. A year passed before the PRR finished this section of the line and no longer had to depend on the antiquated portage railroad.[47]

In 1849, Thomson began organizing the operating department of the PRR. He appointed Herman Haupt, a West Point–trained engineer, as superintendent of transportation.[48] Haupt and Thomson set to work preparing operating guidelines for running trains along the finished portions of the line. Like other railroads of the era, the PRR used timetables and strict operating rules to coordinate train movements.[49]

The Pennsylvania's route paralleled a preexisting commercial telegraph line. In 1846, Henry O'Rielly had built a rudimentary Morse telegraph line across the Alleghenies to connect Pittsburgh with the Eastern Seaboard. Two years later, his company reincorporated as the Atlantic & Ohio Telegraph Company (A&O).[50] Following the reorganization, James K. Moorhead, president of the A&O, approached the board of managers of the PRR seeking permission to move the wires of the A&O to the railroad's right-of-way. In return, he guaranteed access to the commercial telegraph company's facilities. The railroad's directors approved Moorhead's request, provided he adhered to the rules and regulations of the company. The A&O quickly went to work rebuilding their line west of Harrisburg.[51]

In November of 1849, Moorhead once again wrote to the board, perhaps attempting to clarify the terms of the arrangement he had made the previous March. In exchange for permitting the telegraph company to build a line along the whole length of the railroad's right-of-way, Moorhead offered "to do all the telegraphing for the railroad company on the business of the company free."[52] Samuel Merrick, then president of the PRR, accepted Moorhead's terms on the condition that he give the railroad priority use of the wires when required for commercial dispatches. This arrangement provided PRR officials with a convenient means for exchanging business communication quickly along the length of the railroad. Despite the guarantee of priority, nearly a year and a half passed before the company's managers began to use the commercial telegraph line for train dispatching purposes.

The first hint that PRR managers started using telegraphy for operational purposes appeared in the company's January 1852 annual report. Superintendent Haupt remarked that officials were using the A&O line to monitor train movements during daylight hours.[53] Haupt's statement indicates that PRR officials began experimenting with telegraphic train management within a month or two of Minot's successful 1851 test on the Erie Railroad. Unlike the Erie, however, the Pennsylvania did not own the telegraph line along its right-of-way. It depended on A&O operators for service along much of the route. More importantly, PRR officials could use the telegraph line only when it was not occupied with commercial traffic.[54] Consequently, when the A&O was busy transmitting commercial messages, or when they closed the line for the night, the Pennsylvania's managers and train crews simply reverted back to traditional time- and rule-based operating practices.

The Pennsylvania Railroad's growing dissatisfaction with the Atlantic & Ohio's service is reflected in the railroad's corporate minutes. In April 1852, the board consulted with the PRR's solicitor regarding the legal liability of the A&O for an "erroneous communication transmitted by telegraph whereby a collision of trains was produced."[55] The Pennsylvania's counsel reported that he believed the railroad had a legal case against the telegraph company. The board authorized Thomson—who had become company president earlier that year—to pursue the case if he deemed it appropriate, though the company's records do not indicate whether he attempted to recover damages from the A&O.[56]

The deteriorating relationship between the two companies may have motivated the Pennsylvania's board to consider building its own telegraph line. In December 1852, the board created a special three-man Committee on Telegraph to investigate the subject and determine the best course of action for the railroad.[57] The following spring, the committee presented a lengthy report on the subject. During their four-month investigation, committee members had examined the relative merits of the Morse and House telegraph patents. The committee concluded, "The advantages of the House over the Morse lines are . . . so preeminent [that the members of the committee] have no hesitation in recommending it for adoption by the Board."[58] This conclusion stood decidedly at odds with the Pennsylvania's earlier use of the A&O's Morse line.

The Committee on Telegraph advocated the House device for a number of reasons. They found House's design easier for unskilled operators and station agents to use, noting, "The Morse line requires a thorough knowledge of the cipher used [and] constant practice on the part of the operators [and] in case the operator should be absent no one can use or understand it."[59] Accordingly, the committee recommended installing House machines in major stations along the line of the railroad from Philadelphia to Pittsburgh. They also recommended that trains carry portable House machines that could be used to contact stations in the event of breakdowns or delays. Finally, the committee proposed that each station be equipped with telegraph-activated bells that would wake the station agent and alert him to trouble on the line at night.[60]

The telegraph committee's report concluded by reemphasizing the PRR's need for an "independent line of telegraphs," despite the costs involved. They suggested charging the A&O an annual stipend for the firm's use of the PRR's right-of-way, since the commercial company would no longer be providing free service to the railroad. Similarly, the committee proposed leasing poles to the House patent holders in Pennsylvania for a commercial line to compete against the A&O. The committee believed these proposals would "yield to the company an annual amount nearly or quite equal to the interest upon the whole outlay proposed to be made [for the House line]."[61]

The PRR's board of managers considered the telegraph committee's report and authorized the three-man committee to proceed accordingly. Within a few months, however, telegraph committee members were beginning to doubt their initial confidence in the House telegraph. In a September report to the board, the committee admitted that they were thoroughly confused about the relative merits and costs of the competing Morse and House designs. Agents for each side had given them documents "at such direct variance with each other" that the committee recommended placing the whole matter in "some competent scientific hands for the purpose of a thorough investigation." Accordingly, the committee authorized the PRR's vice president to hire an expert who could "make a thorough investigation of the respective merits of the Morse [and] House line of telegraph [and] their adaptation to Rail Road purposes."[62]

In January of 1854, the Committee on Telegraph concluded that Morse's device best fit the railroad's needs. However, they scaled back the length of the railroad's proposed telegraph line. They recommended building a telegraph line only along the most rugged and remote part of the company's right-of-way, between Pittsburgh and Altoona. The committee recommended making a temporary arrangement with the Atlantic & Ohio for a dedicated private telegraph line east of Altoona. Despite these findings, the committee made no firm recommendation to the board. Instead, they asked that the issue be left open for future consideration.[63]

The directors did not make a final decision on the railroad telegraph line until September of 1855. In that year, the PRR's telegraph committee made arrangements to build a Morse line between Philadelphia and Pittsburgh "under the exclusive control of [the railroad] company."[64] Shortly thereafter, the board hired David Brooks, the former superintendent of the A&O's lines in Pittsburgh, as the railroad's new telegraph superintendent. Brooks possessed years of experience as a commercial manager and an in-depth knowledge of Morse telegraphy. In his new position, Brooks slowly began improving the railroad's telegraph infrastructure.[65]

The Pennsylvania Railroad's four-year struggle to determine which telegraph patent to license and then complete a line bore little resemblance to Minot's determined efforts to build a Morse line along the Erie's right-of-way in 1850. While the two companies had begun using telegraphy for train control within months of each other, Minot's early commitment to Morse telegraphy and his desire to institute specific standards for telegraphic dispatching on the Erie put the New York railroad far in advance of the PRR with respect to operational management.

Conceptualizing Telegraphic Train Management

By the time the Pennsylvania Railroad finally constructed its own Morse line, Minot's eventual successor as Erie general superintendent, Daniel C. McCallum, had begun developing a telegraphic management framework that would serve as a model for American railroad officials through the end of the Civil War era. McCallum was a vocal advocate of railroad telegraphy. Like British telegraph pioneer William F. Cooke, he argued that telegraphic train management would save American railroad firms money and promote safety and efficiency on the country's rapidly expanding rail network. However, McCallum's brash and grasping personality, and his shortcomings as a personnel manager, prevented his highly innovative management protocols from being effectively implemented on the Erie in the antebellum era.[66]

In 1854, the Erie's board of directors asked Minot to implement a set of strict employee rules devised by McCallum, at that time one of the railroad's divisional superintendents. Board members felt that Minot's management style was much too lax. They were concerned that his close relationship with the railroad's workforce prevented him from making painful but necessary decisions about wages and working conditions. By contrast, McCallum's proposed rules promoted employee discipline by limiting locomotive engineers' freedom to operate their trains as they saw fit. Board members believed that a more efficient workforce would save the struggling company money. Despite pressure from the company's directors, Minot refused to implement the reforms, and McCallum took his place as general superintendent.[67]

McCallum had begun working for the New York & Erie Railroad in 1848 as a civil engineer. In October 1852, he was promoted to superintendent of the Erie's western Susquehanna Division.[68] As a division superintendent, McCallum focused on systematizing the Erie's operating rules for train crews and imposing order on the struggling railroad.[69] After the board appointed him general superintendent in 1854, McCallum began imposing his new rules across the entire Erie network. Erie train crews pushed back against the new rules, as well as McCallum's policy of blackballing fired locomotive engineers by sharing their names with managers from other railroads.[70] When McCallum refused to do away with the new standards, Erie engineers went on strike, the first in the railroad's history. After a ten-day work stoppage that completely paralyzed traffic along the rails, McCallum was forced to compromise with the striking workers.[71]

McCallum also directed his energies toward improving the operational efficiency of the Erie, particularly within its telegraph department. Shortly after the strike ended, McCallum issued a special employee rule book for telegraphers. This was the first telegraph rule book issued on an American railroad. The book laid out clear guidelines for the transmission and reception of messages, for record keeping, and for operator etiquette. McCallum required operators to follow standardized procedures when issuing train orders from divisional superintendents to train crews. All train orders had to be "signed by the Conductor personally" before they could take effect.[72] This rule forced conductors to pay closer attention to the new orders they had received. Station operators also had to employ a series of standardized, alphanumeric codes when sending messages. For example, when requesting a train location on the rail line, operators had to use code 19—"Have you any report of . . . ?" Reports on train locations required code 33—"Train left this Station at . . . ." Finally, McCallum assigned each Erie railroad station a two-letter abbreviation to simplify message transmission.[73]

Along with standardizing telegraphic communication practices on the Erie, McCallum devised standardized train dispatching forms for divisional superintendents. "Blank 31" and "Blank 32" corresponded to the aforementioned number codes used by telegraph operators.[74] Orders issued on Blank 31 required a train crew to pass a train going the opposite direction at a specified location, regardless of what the daily timetable dictated. Train engineers and conductors had to acknowledge the receipt of the order at the telegraph office and sign the corresponding form. The operator then had to reply using the corresponding code and repeat the entire message back to the superintendent to double-check the message for errors. The orders were not valid until all train crews affected by them had acknowledged their receipt.[75]

The other form that McCallum instituted came to be known as "Blank A." Blank A train orders typically involved minor changes to a train's schedule, and the operator did not have to repeat the order back to the superintendent for verification. More importantly, train crews did not have to stop their trains to sign and confirm the orders, as they did with a Blank 31 order. McCallum had Blank A forms printed and distributed to telegraph operators in each station. The preprinted forms saved additional time, since operators had to write only the initial and the final destinations for each train on the form and deliver it to the intended train crew as they passed by the station.[76] The Erie continued using McCallum's train dispatching practices with few modifications until the late 1870s. They also provided the basis for the standard train order forms used by other railroads during the Civil War era.[77]

Along with his practical improvements to the Erie's train dispatching framework, McCallum championed the principle of telegraphic train management in his "Superintendents Report," which appeared in the New York & Erie's *Annual Report* for 1855.[78] The Erie's traffic had increased dramatically since it opened at the beginning of the decade. The company operated approximately 200 locomotives and owned over 2,700 freight cars and 170 passenger cars.[79] In his report, McCallum noted that long-distance railroads like the Erie required a different form of administration than "a [rail]road fifty miles in length."[80] He called for "a system perfect in its details, properly adapted and vigilantly enforced" to reduce costs and improve efficiency. Telegraphic train dispatching formed the core of McCallum's management model. He noted that without the telegraph, "business could not be conducted with anything like the same degree of economy, safety, regularity, or dispatch."[81] As his report made clear, McCallum viewed these four principles as necessary for proper railroad management.

McCallum continued his 1855 report by describing the telegraphic reporting practices that he had instituted during his two years as general superintendent:

> Hourly reports are received by telegraph, giving the position of all the passenger and the principal freight trains. In all cases where passenger trains are more than ten minutes, or freight trains more than half-an-hour behind time, on their arrival at a station the conductors are required to report the cause to the operator, who transmits the same by telegraph to the General Superintendent; and the information being entered as fast as received, on a convenient tabular form, show at a glance, the position and progress of trains in both directions on every Division of the [rail]road.[82]

The administrative reporting framework enabled officials to identify and respond quickly to delays by issuing new train orders. It also allowed McCallum to identify and address the more systemic causes of the delays and prevent them from reoccurring.[83]

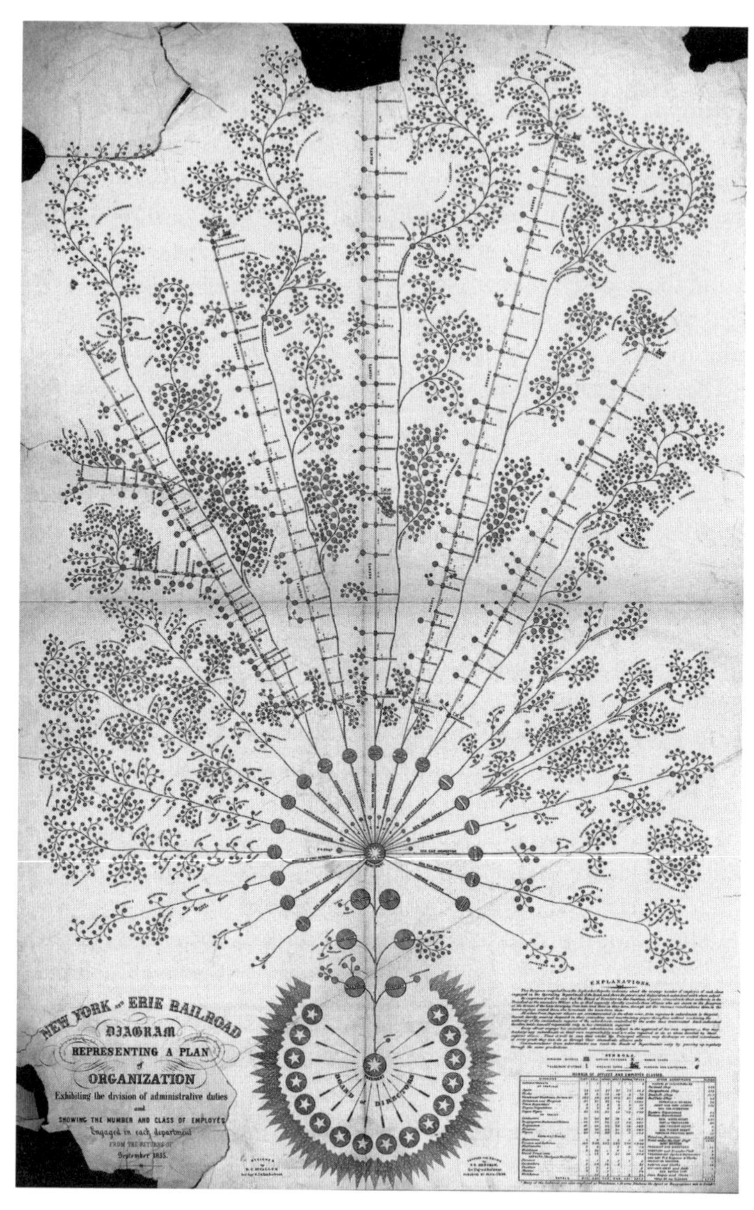

New York & Erie Railroad Superintendent Daniel C. McCallum devised the earliest known example of a modern organizational chart. The diagram appeared in his 1855 *Annual Report* and illustrated the lines of managerial authority on the railroad. Unlike later organizational charts that represented hierarchical lines of authority in a pyramidal manner, McCallum's chart showed the lines of authority radiating out from a central core, much like branches and leaves from a tree trunk. *Library of Congress, Geography and Map Division*

McCallum's sophisticated management model garnered him attention from the railroad press. The *American Railroad Journal* followed his managerial innovations throughout the mid-1850s. The journal's editors widely promoted McCallum's statement that he "would rather have a [rail]road of a single-track with the electric telegraph to manage the movement of trains, than a double-track without it."[84] This comment mirrored the views of Cooke and other European telegraph promoters. McCallum's telegraphic train reporting practices implemented the "bird's-eye" model of railroad management that Cooke had championed in the early 1840s.[85] McCallum's widely publicized managerial practices also provided a model for other American railroad officials to examine as they contemplated the role of telegraphy in railroad operations.[86]

McCallum's poor understanding of labor relations ultimately proved to be his undoing as general superintendent. Despite his compromise with striking Erie employees in 1854, locomotive engineers resented his harsh policies and cost-cutting practices. In 1856, 150 engineers went on strike to protest the arbitrary dismissal of a fellow employee. The strike lasted six months and nearly ruined the company. Unwilling to compromise with the strikers, McCallum tendered his resignation in March of 1857.[87]

McCallum's brief foray into intensive telegraphic railroad management represented an anomaly, rather than the beginning of a broader trend in the antebellum railroad industry. After he left the Erie, the financial Panic of 1857 forced Erie officials to severely curtail operating expenses and discontinue many of the telegraphic management practices McCallum had developed for the railroad. An employee rule book issued in August 1857 illustrated the new status quo. It did not contain any of McCallum's complex procedures for managing train movements. Instead, it relegated telegraphic dispatching to the category of "Movement of Trains By Special Order." This change demonstrates that senior Erie officials did not see the telegraph as an essential component of railroad management at a time when spare capital was desperately needed to make up for operating losses. Once again, timetables and detailed operating rules would serve as the primary basis for daily train operations on the Erie.[88]

Resistance and Adoption in the East

By the middle of the 1850s, the Erie and the Pennsylvania Railroads stood at the technical and managerial forefront of American railroads. Both companies had instituted some form of telegraphic train management early in the decade. Under Minot and McCallum's oversight, the Erie closely integrated telegraphy with operational management. PRR officials gradually developed their own telegraphic management practices, though they were not as advanced as the Erie's. Managers with other eastern railroads, however, displayed a much more ambiguous relationship with telegraphy. A few long-distance rail lines accepted it as a helpful supplement to traditional management practices, while other companies strongly resisted telegraphy and viewed it as a threat to safe railroad operations.

In 1853, the Baltimore & Ohio Railroad completed the third trans-Appalachian rail line.[89] While the railroad was still under construction in 1848, telegraph promoters organized the Western Telegraph Company to connect Baltimore with the Ohio River Valley. The commercial managers chose to run their line along the National Road, rather than approach the B&O about the possibility of using its existing right-of-way. Neither firm had any dealings with the other until June 1853, six months after the B&O had completed its main line from Baltimore to Wheeling, Virginia, when B&O directors entered into an agreement with the Western Telegraph Company for access to its line. Railroad officials intended to use the commercial line for monitoring train locations. The agreement provided for limited connections between the Western's line and a few key stations along the B&O's route. Two years later, B&O officials agreed to pay the telegraph company $25,000 to move its line onto the railroad's right-of-way. The new agreement hinted that the company was beginning to make greater use of the telegraph for dispatching purposes. Still, B&O officials chose not to invest in their own railroad telegraph network before the Civil War.[90]

A similar situation took place on the young New York Central & Hudson River Railroad (NYCR). The New York, Albany & Buffalo Telegraph Company line paralleled the tracks of many of the smaller railroad companies that were merged to form the NYCR in 1853. Prior to the merger, the managers of the individual companies had derided the telegraph company and expressed no interest in using its line for railroad business.[91] After the merger, however, the new company's directors approached the NYA&B about leasing a telegraph line for railroad business. The telegraph company's president offered the railroad a liberal contract, but the board members rejected it as being too generous.[92] A year passed before the railroad and the telegraph company reached a new agreement. The NYCR agreed to give the NYA&B $10,000 in bonds in exchange "for the exclusive use, for railroad purposes, by this company, of one of the wires of said telegraph [company].[93]

NYCR managers soon grew dissatisfied with the service provided by the NYA&B. By 1856, officials complained that the railroad telegraph line no longer functioned, and they began withholding lease payments. Telegraph officials quickly sent two foremen along the entire line repairing connections and cleaning the insulation. Ten days later, it was put back into service. As J. D. Reid commented in his telegraph history, "The railroad officers at once noticed the change, and were highly pleased."[94] This incident reveals that NYCR officials did not depend on telegraphy for railroad operations in 1856, since deteriorating telegraph service had not significantly impacted overall train service. While NYCR officials had come to view the telegraph as a useful supplement for traditional operating practices, they did not consider it a necessary management tool.

Shorter railroads in the East as well did not follow the lead of the long-distance early adopters. Since the early 1850s, a few telegraph promoters had tried to sell the idea of telegraphic management to regional rail lines, primarily as a strategy for gaining access to their rights-of-way. In 1852, American telegraph promoter Henry O'Rielly issued a circular echoing British telegraph pioneer Cooke's argument that telegraphy would increase both the safety and efficiency of American railroads at a nominal cost: "A well-arranged telegraph for railroad purposes would, each and every year, render to a railroad company sufficient benefits to counterbalance the whole cost of construction."[95] O'Rielly's financial logic did not resonate with many American railroad officials on short lines. As a British reporter traveling in the United States two years later noted, "The telegraph is rarely seen in America running beside the railway, for what reason we do not know."[96] Had the British reporter made a systematic survey of smaller rail firms in the East, he would have found that most officials saw little reason to employ expensive telegraphic train management practices on their lightly trafficked rail lines in the mid-1850s.

The well-capitalized Philadelphia & Reading Railroad (Reading), which primarily transported anthracite from the coal mines of eastern Pennsylvania to Philadelphia, was an exception to this broader trend. The Reading began managing train operations by telegraph in 1855.[97] It also had been one of the earliest railroads in the country to allow a commercial telegraph line along its right-of-way. After the commercial line was completed in 1847, the Reading leased access to the telegraph line for business correspondence.[98] By 1855, Reading managers were dispatching trains telegraphically twenty-four hours a day during the peak coal season. The same year, Gustavus A. Nicolls, the Reading's superintendent, urgently requested a separate telegraph line for transmitting train orders because heavy commercial business delayed train dispatches and other pressing operational business. Since delayed train orders could back up traffic along the entire Reading line, time was of the essence. For this small but busy anthracite railroad, telegraphy represented an invaluable management tool.[99]

By contrast, officials on regional rail lines in New England continued to shy away from incorporating telegraphy into operational management in the 1850s. The response of New England railroad managers to a major collision on the Providence & Worcester Railroad (P&W) in 1853 illustrates their resistance to telegraphy. In August of that year, two passenger trains collided head-on near the town of Valley Falls, Rhode Island.[100] Subsequent investigations assigned blame to the untrained conductor who had been running one of the trains.[101] Unaware that his watch was running slow compared to the P&W's official time, the conductor believed that his delayed train had enough time to cross onto the double-track portion of the P&W's line before a northbound train entered the single-track section of the line.

The crash killed fourteen passengers and led to an investigation by the railroad commission for the state of Rhode Island. The inquiry revealed that the P&W's timetable allowed as little as one minute for trains to reach sidings before opposing trains passed them. Additionally, the company made no effort to check the accuracy of train crews' watches before they started their daily runs. Based on this evidence, the commissioners recommended that the Rhode Island legislature consider stricter regulation of the state's railroads. The P&W tried to appease the railroad commissioners and avoid additional legislative oversight by dismissing the train crews involved in the crash and offering to double-track more of the company's line.[102]

Other New England railroads responded to the Rhode Island crash in two different ways. The Boston & Providence managers instituted stricter standards for employee timepieces. Their goal was to eliminate unsynchronized watches as a possible cause of collisions. Some firms increased the mandatory time interval for trains passing on single-track lines, which allowed delayed trains additional time to pull into sidings before meeting opposing traffic.[103]

New England railroad managers did not view telegraphic train dispatching as a realistic solution to the dangers of operating two-way traffic on single-track lines. Although telegraphy would enable managers to locate trains precisely and to safely control their movements, officials instead focused on time- and rule-based solutions. In the case of the P&W, company officials ultimately chose to double-track problematic sections of the rail line, an expensive option compared to the cost of telegraph service.[104] New England managers continued to employ incremental technical fixes to address increasingly complicated operational challenges, rather than make use of telegraphy in any form.

Limited Adoption in the Midwest

While New England railroad managers lagged behind mid-Atlantic trunk line managers in their use of telegraphy in the antebellum period, the Midwest emerged as an innovative locus where railroad officials and telegraph promoters increasingly found common ground. A handful of telegraph promoters owned the major telegraph firms in the region. As these businessmen attempted to consolidate their control over key routes between midwestern cities in the mid-1850s, they looked to railroad companies as partners. The activities of the managers of the Caton lines in Illinois, and the Western Union lines in Ohio, Indiana, and Illinois, illustrate telegraph promoters' growing demand for access to railroad rights-of-way in the mid-1850s. However, they also demonstrate that midwestern railroad managers did not necessarily adopt telegraphic train management practices more rapidly or consistently than their eastern brethren, despite their greater willingness to permit telegraph lines on their property.

Illinois quickly emerged as an important center for telegraph and railroad construction at the end of the 1840s. In the North, the young city of Chicago expanded rapidly into an important commercial center and transshipment point for agricultural products from the surrounding region. The state became a crossroads for north-south and east-west transportation and trade. In 1849, telegraph promoters organized the Illinois and Mississippi Telegraph Company (I&M) to connect St. Louis with Chicago. The promoters elected Judge John D. Caton, a member of the Illinois Supreme Court, as one of the company's directors.[105] Caton soon became actively involved in the management of the company and even acquired some proficiency as an operator. In 1852, he was elected president of the I&M lines and began negotiating right-of-way contracts with regional railroads.[106]

Caton signed his first contract with the Illinois Central Railroad (IC). In 1851, investors had organized the IC to connect Chicago with Cairo, located at the confluence of the Ohio and Mississippi Rivers at the southern end of the state. The first section of the IC's line was completed in 1853. Later in the year, the company's directors made arrangements with Caton to build a telegraph line along the IC's right-of-way.[107] He personally acquired Morse licenses and incorporated the Illinois Central Telegraph Company in January of 1854.[108] As the railroad continued building its line the following year, the IC Telegraph Company extended the telegraph line at the request of the IC's board.[109] By the spring of 1856, the railroad and telegraph lines spanned the entire length of the state from Chicago to Cairo.[110]

Caton's contract with the IC provided the railroad company with free use of the line, and the railroad in turn pledged to protect the property of the telegraph company, provide operators for each railroad station, and transport telegraph company employees, equipment, and construction supplies free of charge.[111] Moreover, Caton secured exclusive access to the railroad's right-of-way. This privilege enabled him to quickly enlarge his commercial telegraph network, while blocking the expansion efforts of rivals.

As he was completing his Illinois Central telegraph line, Caton entered into an agreement to build a line for the Chicago, Burlington & Quincy Railroad (CB&Q). In 1856, the CB&Q's directors authorized him to form the Chicago and Mississippi Telegraph Company and construct a telegraph line along the railroad's right-of-way from Chicago to Burlington, Iowa.[112] Within months, Caton brought both the Illinois Central and the Chicago and Mississippi Telegraph Companies under the umbrella of the Illinois and Mississippi Telegraph Company.[113]

While both the IC and the CB&Q had telegraph lines along their rights-of-way by late 1856, neither company used their lines for managing train operations, except in extraordinary circumstances. Illinois Central employees depended on traditional timetables and rule books during the late 1850s. IC rule books for 1857, for instance, contain no mention of the telegraph. A survey of corporate correspondence prior to 1863 reveals only a handful of instances in which the telegraph was used to issue train orders.[114] A similar situation prevailed on the CB&Q. Company records contain no references to telegraphic train control until the Civil War period. Only in 1863 did the company's directors enter into an agreement with Caton's Illinois and Mississippi Telegraph Company for a second telegraph line "for the exclusive use, management, and control of the [railroad company]."[115]

Caton found other Illinois railroad managers somewhat more receptive to employing telegraphy in an operational role. In 1856, he signed a contract with the Chicago, Alton & St. Louis Railroad (Chicago & Alton). The railroad's directors wanted telegraph service along their line from St. Louis to Joliet, Illinois. Following the completion of the telegraph line, Caton arranged for round-the-clock service. He appointed Marvin Hughett, a former telegrapher for the NYA&B line, as the head operator. Hughett also served as one of the first train dispatchers in the country. He was granted express authority from senior officials to use the telegraph to monitor train locations and to issue any necessary orders to train crews. At stations without twenty-four-hour service, night watchmen had to notify Hughett if trains were running more than fifteen minutes behind schedule. Reid later reported that during Hughett's administration, "not a car was scratched by collision, the position of every train was exactly known, and the greatest confidence inspired."[116]

Other telegraph promoters worked to secure valuable railroad concessions in the neighboring states of Indiana, Ohio, and Michigan. Jeptha H. Wade built the first telegraph line west of Buffalo along the tracks of the Michigan Central Railroad in the late 1840s despite resistance from the railroad's president, John W. Brooks. A few years later, he extended the line southwest through Ohio and Indiana to St. Louis, Missouri. Like Caton, Wade saw railroad rights-of-way as strategic locations for telegraph lines. In 1854, he urged his colleagues at the struggling Cleveland & Cincinnati Telegraph Company to improve and expand their original commercial line as quickly as possible, and "when suitable arrangements can be made . . . get it on to Railroads."[117]

In 1856, Ezra Cornell, Caton, Wade, and a number of other telegraph promoters founded the Western Union Telegraph Company in New York. The new company united the financial interests of approximately half a dozen major midwestern telegraph networks and began absorbing other companies throughout the United States.[118] Acting as an agent for the new firm, Wade acquired control of the decrepit telegraph line that he had originally built along the Michigan Central's tracks. He met with Brooks to sign a new Western Union construction contract. Wade was unsure whether Brooks would order the old line removed from his right-of-way or give Western Union a chance to demonstrate the usefulness of a new line.[119] Brooks finally gave his approval after being assured by Wade that the new line "could be made useful and satisfactory to him."[120]

The Michigan Central's 1856 annual report reflected Brooks's change of opinion about the telegraph. He praised it for rendering operations on the firm's single-track line "efficient, reliable, and certain." Both he and his superintendent, Reuben N. Rice, discussed its value for daily operations; however, the two men also observed that the telegraph was of greater value to railroad management when it was under the exclusive control of the railroad. Railroad messages would not have to compete with commercial traffic, and the line could be kept open twenty-four hours a day when necessary. Consequently, the Michigan Central began building its own independent telegraph line. Brooks's complaint would become widespread among railroad officials in the post–Civil War era as they chafed under the constraint of conducting railroad business using telegraph lines owned and operated by Western Union and other commercial telegraph firms.[121]

Wade and his close associate Anson Stager, who served as Western Union's general superintendent, made great efforts to secure exclusive contracts with other midwestern railroads during the latter half of the 1850s. In 1856, Wade obtained access to the right-of-way of the Michigan Southern Railroad (later the Lake Shore & Michigan Southern, an affiliate of the New York Central), the company Charles Minot worked for after leaving the Erie in 1854. The next year, Wade took over the Atlantic & Ohio Telegraph Company, whose poles ran along the Pennsylvania Railroad's right-of-way. By the end of the decade, Western Union had established exclusive control over the major railroad rights-of-way in the Midwest. In exchange, the company agreed to provide railroad firms with free access to its telegraph lines for sending corporate correspondence or for transmitting train orders.[122]

Despite the free access clause in Western Union contracts, Wade's exclusive agreements with railroad firms served the interests of Western Union far more than those of the railroad companies. While a few midwestern railroad officials began experimenting with telegraphic dispatching by using the free telegraph service provided by Western Union, the majority continued to view telegraphy with distrust. As Reid recalled: "Even to the alert western mind, it was a work of some difficulty to prove the value of the telegraph in moving and directing trains. It seemed like devolving on mechanism the gravest responsibilities connected with the safety of human lives."[123] For conservative railroad officials in the region, strict time- and rule-based operating practices, not telegraphy, provided the best protection against accidents and employee mistakes.

Conclusion

Technical and organizational change was inconsistent and nonlinear during the decade of the 1850s. Erie and Pennsylvania Railroad managers pioneered techniques for utilizing the telegraph to oversee rail operations at the beginning of the decade. After ten years, however, senior PRR officials still emphasized traditional operational management practices and placed tight restrictions on how the telegraph could be employed for reporting train locations and issuing orders.[124] Similarly, financial factors influenced Erie Railroad officials' decision to de-emphasize telegraphic management following McCallum's 1857 resignation. Even though train traffic moved around the clock on both lines by the end of the decade, the PRR and the Erie continued to close many of their telegraph offices at night until the beginning of the Civil War.[125]

Managers of regional railroads with fewer economic resources than large trunk lines like the PRR, Erie, and Michigan Central (or busy regional lines such as the Philadelphia & Reading) had no compelling need to adopt telegraphic train management. Many of these lines did not handle heavy volumes of traffic in the 1850s. For these firms, daily train movements might consist of no more than a dozen trains, and these could be managed through simple timetables and operating rules. While numerous railroad officials signed right-of-way agreements with Western Union agents in the late 1850s, in exchange for free telegraph service, few used the service for monitoring train movements or issuing instruction to train crews. This suggests that many railroad officials found the telegraph extremely useful as a communication tool for business correspondence, but not as a dedicated management tool for monitoring and guiding train movements.

At the end of the 1850s, most railroad officials continued to believe that strict time- and rule-based operating practices, not telegraphy, best protected their firms from collisions and other routine operating accidents. They did not view telegraphy as a tool necessary for operational management, even though many increasingly used the telegraph for commercial correspondence. While managers occasionally employed telegraphy for train dispatching in extraordinary circumstances, many continued to treat it as a dangerous expedient. Instead, incremental improvements in operating rules, train scheduling, and timekeeping satisfied the operating requirements of most officials. The telegraph remained marginalized until an overwhelming volume of military and civilian traffic during the Civil War forced railroad managers to utilize it to keep traffic flowing smoothly on their rail lines.

At War with Time and Space

> When nine-tenths of the work done by the telegraph is railroad business and often its promptness and correctness are of vital consequence, it would seem proper for the railroad managers to control the men whose services are so important to them.
>
> Daniel C. McCallum, Director, U.S. Military Railroad, 1864

During the opening days of the Civil War, Simon Cameron, the newly appointed United States secretary of war, desperately sought assistance from loyal railroad and telegraph managers in the North and Midwest. Cameron recognized that the railroad and telegraph would play an important role in the coming hostilities, but career US military officials did not have practical experience building or managing railroad or telegraph lines.[1] During the country's previous major conflict, the Mexican War of 1846, military officials used wagon trains to transport supplies and ammunition. Horseback couriers provided critical communication services, and troops marched toward the battlefield on foot. Following the Mexican War, civilians developed increasingly sophisticated railroad and telegraph networks in many parts of the East and Midwest, but in the early days of the Civil War, military officials were not prepared to use the new infrastructure to support troops in the field. As a result, Cameron looked to civilian railroad and telegraph officials to oversee the rapid development of a complex military railroad and telegraph network during the conflict.

The federal government's network of wartime railroad and telegraph lines was unprecedented in its size and complexity. The United States Military Railroad (USMRR) stretched over twenty-one hundred miles throughout the South, a distance nearly equivalent to the combined mileage of all northern railroad firms at the beginning of the Civil War, and employed twenty-five thousand civilian employees at its peak. Its sister organization, the Military Telegraph Corps, operated a network of telegraph lines and cables that extended fifteen thousand miles and employed roughly one thousand civilian operators, in addition to countless linemen and manual laborers.[2]

Civilians managed these two unwieldy networks based on familiar prewar operating practices. At the beginning of the conflict, Thomas A. Scott, Andrew Carnegie, and other Pennsylvania Railroad (PRR) officers who answered Secretary Cameron's appeal for assistance established a closely integrated military railroad and telegraph network modeled after the PRR's prewar operating structure.[3] A year later, military authorities separated railroad and telegraph operations into two independent departments under the direct oversight of the secretary of war. Most civilian officials did not object to the new division between the telegraph and railroad departments. A few like Daniel C. McCallum, formerly of the Erie Railroad, found the split to be highly problematic. As a civilian railroad manager, he had come to depend on unfettered access to the telegraph for effectively monitoring and controlling train movements along the length of the railroad lines under his control.[4] McCallum's and his cohort's vocal opposition to losing control over telegraph facilities along military rail lines revealed a growing schism between traditional railroad management practices and newer styles of operational management predicated on instantaneous communication and ready access to information on train locations and movements.

The crucible of wartime operations provided a stage on which the conflicting managerial prerogatives of civilian railroad authorities played out over the course of the war. As military railroad and telegraph lines spread throughout the theaters of battle, they provided concrete examples of how telegraphy could be used to ensure safe and timely train movements across the expansive network. Railroad and telegraph officials gradually recognized that coordinating their activities, instead of competing for power and resources, was necessary for guaranteeing safe and effective rail operations. This, in turn, helped promote acceptance of telegraphic train management practices by civilian railroad managers and officials during and after the war.[5]

Crisis of Leadership

On April 19, 1861, less than a week after Confederate forces began the war by shelling Fort Sumter, an angry crowd of prosecession Baltimore citizens blocked the passage of two Massachusetts infantry regiments traveling by train from Philadelphia to Washington, DC. The rail lines entering Baltimore from the north and south did not connect in the city, so the soldiers were in the process of marching along the docks from the Philadelphia, Wilmington & Baltimore's President Street Station to the Baltimore & Ohio's (B&O) Camden Station when the mob attacked.[6] Four soldiers were killed and numerous others were wounded in the ensuing Pratt Street Riot. Anxious to bolster the small number of loyal troops defending the capital, but unwilling to further antagonize the citizens of Baltimore, President Abraham Lincoln sought an alternative transportation route for men and supplies.

General Benjamin F. Butler, the commanding officer of the Massachusetts volunteer regiments, located an old steam ferry approximately fifty miles north of Baltimore at Havre de Grace, Maryland, near the mouth of the Susquehanna River. He used the ferry to transport his troops down the Chesapeake Bay to

Annapolis. From there, Butler took control of the Annapolis & Elk Ridge Railroad, which connected with the B&O's Washington Branch a few miles south of Baltimore.

Andrew Carnegie accompanied the Massachusetts volunteers to Washington at the request of Thomas A. Scott, his former boss at the Pennsylvania Railroad. In early April, Scott had contacted Carnegie at the behest of Secretary of War Cameron and asked him to volunteer his services to the War Department as an expert telegrapher and railroad manager. Without hesitation, Carnegie set out for Washington.

As Carnegie's train proceeded through the Maryland countryside, he noticed that the telegraph wires along the track had been staked to the ground, shorting the electrical circuit and rendering the lines unusable. He quickly dismounted and pulled out the stakes, which, "when released, in their spring upwards," Carnegie later wrote, "struck me in the face, knocked me over, and cut a gash in my cheek which bled profusely. In this condition, I entered the city of Washington with the first troops."[7] While Carnegie likely embellished his exploits in the opening days of the war, he did provide invaluable service as one of the first officials in the federal government's rudimentary railroad and telegraph department.

Scott's request to Carnegie came at a time of great uncertainty for the North's military leadership, particularly those overseeing the United States Army's transportation and communication infrastructure. Secretary of War Cameron took office barely a week after Lincoln's inauguration in March. A major railroad investor and speculator, as well as former Pennsylvania senator, Cameron spent his first few weeks granting favors to political supporters rather than focusing on the impending secession crisis. In the meantime, many senior United States Army officers resigned their commissions and returned to their ancestral homes in the South. In particular, General Joseph E. Johnston, the respected and capable head of the US Army Quartermaster Department, left the army late in April 1861, a week after the president called for seventy-five thousand state militia to suppress the Southern rebellion.[8]

The sudden exit of senior logistical experts from the Quartermaster Department, which oversaw the supply and transportation of military forces, left Cameron in a tenuous position. The career officers who remained in the Quartermaster Department knew little about railroad operations and electrical telegraphy. Seeking advice and assistance, Cameron turned to his old friend and business associate J. Edgar Thomson, president of the Pennsylvania Railroad.[9] Thomson recognized the serious difficulties involved in moving troops and supplies to the Northern capital. The only rail line to Washington originated in Baltimore. Southern sympathizers could easily block this route by destroying bridges or removing sections of rail, unless troops guarded the entire length of the line, which was infeasible.

The riots in downtown Baltimore on April 19 confirmed Thomson's assessment of the District of Columbia's vulnerability. In addition to blocking rail lines, Baltimore city leaders had cut the long-distance telegraph lines passing through the city, further isolating the Northern capital from contact with surrounding states.[10] Like other slaveholding border states, Maryland was politically divided into Unionist and prosecession sympathizers. Some Maryland politicians wanted the state to join the Confederacy. Thomson, like many other observers, realized that Maryland must remain in the Union, and the rail link from Baltimore to Washington must stay open, or else the District of Columbia would have to be abandoned as the federal capital, which would be a significant symbolic loss in the early days of the war.

As hostilities escalated in April, Thomas A. Scott, Thomson's vice president at the Pennsylvania Railroad, relocated to Harrisburg to oversee transportation and communication resources for militia troops being inducted into federal service.[11] During the closing weeks of April, Scott dealt with transportation and communication crises created by the rail and telegraph blockade of the Northern capital. He rerouted troops arriving in Pennsylvania from the West through Philadelphia and Annapolis to Washington, and attempted unsuccessfully to open an alternative telegraph connection with the capital through Hagerstown, Maryland.[12] Scott's tireless efforts showcased his value as a railroad manager to the new and untested secretary of war.

By the end of April, Cameron acknowledged the futility of depending on private companies to operate railroad and telegraph lines in Maryland.[13] He now believed that only military authority could ensure the continued operation of the transportation and communication resources of the unstable region and keep lines of communication open between the capital city and points north.[14] Cameron, however, could not locate anyone serving in the Northern military bureaucracy with the necessary skills to manage railroad and telegraph operations. On April 21, he officially asked Scott to come to Washington and oversee all railroad and telegraph lines controlled by the federal government. Scott arrived in Washington a few days later. Carnegie and other managers from the Pennsylvania Railroad soon joined him.[15]

A State of Confusion

On April 27, 1861, Cameron officially announced that Scott had full authority over all railroad and telegraph operations between Washington and Annapolis, Maryland.[16] Days earlier, Cameron had placed troops in the B&O's depot north of the US Capitol Building and had announced that all train movements between Washington and Annapolis would be under military control. B&O officials in Baltimore objected to Cameron's actions, since he had not consulted with them before making his decision. In response, they withdrew all company equipment and rolling stock from the affiliated Annapolis & Elk Ridge Railroad, leaving the military line with only a few dilapidated locomotives and cars.[17] Cameron pushed back by ordering General Butler to seize control of the B&O's entire Washington Branch line.[18] The roughly forty-mile route formed the first link in the federal government's military railroad network.

For the first nine months of the war, PRR officials staffed many of the senior posts in the growing railroad and telegraph network. Andrew Carnegie served as one of Scott's chief lieutenants. After Scott requested his services in late April, Carnegie contacted David McCargo, superintendent of the Pennsylvania Railroad's telegraph network, and asked for four skilled telegraphers to assist him with managing government railroad and telegraph operations.[19]

When Carnegie arrived in Washington at the end of April, he found "an incredible state of confusion and inefficiency in nearly every department of the Government," as he later recalled to chronicler John Emmet O'Brien.[20] Carnegie quickly set to work assigning new duties to his fellow PRR managers. He attempted to replicate the Pennsylvania Railroad's operational managerial structure, which supplemented time- and rule-based operations with limited telegraphic train dispatching. Carnegie placed David Strouse, the PRR's telegraph agent from Mifflin, Pennsylvania, in charge of the federal government's rudimentary telegraph network. The government network handled both telegrams and train dispatches. However, the institution existed in name only. Strouse had no equipment, construction materials, or federal funding to aid him with his task. Instead, he struck an informal arrangement with Edward S. Sanford, president of the American Telegraph Company (ATC). For the first six months of the war, the ATC provided Strouse with much-needed supplies for connecting War Department buildings with the ATC's long-distance lines, and staff for maintaining and operating the circuits.[21]

With the government's telegraph network in Strouse's capable hands, Carnegie personally took on the challenge of managing military rail operations. His first significant task involved reopening the rail lines through Baltimore. Once General Butler's forces imposed martial law on the city, crews from the Northern Central and Philadelphia, Wilmington & Baltimore Railroads quickly set to work repairing damaged bridges and rails. By the middle of May, commercial train traffic once again moved freely into Baltimore from the north and west. The B&O's Washington Branch, however, remained under military control.[22]

Next, Carnegie and Scott concentrated on connecting the rail lines within the capital region to facilitate troop and supply movements through Washington and across the Potomac River. The B&O line in Washington did not connect with the rail lines running along the Virginia shore, because none of the bridges across the river could support loads heavier than pedestrian and wagon traffic.[23] Scott decided to extend the B&O's Washington Branch south across

Capitol Hill and along the north bank of the Potomac River to the Long Bridge.[24] If the bridge could be reconstructed to handle railroad traffic, a link could be established with a number of strategic rail lines in Virginia. The connection would also enable the US Army to begin stockpiling supplies and equipment on the south bank of the Potomac in anticipation of a march on Richmond, the capital of the Confederacy.

Secretary Cameron provided Scott with funding to begin rebuilding the Long Bridge and laying track. Over the course of a week, laborers working day and night completed work on the bridge and repaired rail lines damaged by Confederate forces. When the bridge was finished, the new rail connection linked the rail lines leading into Alexandria, Virginia, with the B&O's branch to Baltimore.[25]

Shortly after work on the Long Bridge ended, Cameron officially placed Scott in charge of all railroad and telegraph operations controlled by the federal government. He "washed his hands of all direct responsibility in regard to these facilities, asserting that all orders relating to the extension or operation of these [rail]roads must come from Scott."[26] The formal statutory origins of the United States Military Railroad (USMRR) are obscure, most likely owing to the chaotic nature of federal wartime operations in the first months of the war. Congress eventually authorized the USMRR to operate all confiscated southern railroad property in January 1862, but Scott had been running captured rail lines in northern Virginia as "Government Railways" under the auspices of the War Department since Cameron's formal declaration in May of 1861.[27]

As Scott's deputies, Carnegie and Strouse played critical roles in keeping government transportation and communication facilities functioning in the early days of the war. Both men managed entities largely by expediency instead of with deliberate forethought. While these organizations fulfilled their missions, each group evidenced disorganization and significant operational inefficiencies.

Command–Control–Communicate

On May 24, 1861, Confederate forces withdrew from Alexandria and much of the Potomac shore across from Washington. Federal troops soon occupied the strategic river town. Carnegie moved his headquarters from Washington to Alexandria shortly after the federal advance. He took possession of railroad facilities abandoned by Confederate forces and quickly began turning Alexandria into a supply depot for rail operations in northern Virginia.

Carnegie struggled to return the Orange & Alexandria Railroad (O&A) to running condition. Confederate troops had destroyed rolling stock, bridges, and telegraph lines as they retreated westward from Alexandria to Manassas Junction, Virginia, in late May. Carnegie had to beg, borrow, and requisition equipment to make repairs. During the early weeks of June, Carnegie oversaw repair efforts while Scott obtained locomotives and rolling stock from the Philadelphia & Reading Railroad and other northern lines.[28]

Virginia's prewar railroad infrastructure played a major role in shaping military campaigns in the East. No rail lines connected Washington with Richmond because Virginia did not allow railroads chartered within the state to connect with lines in surrounding states. The policy resulted from Virginia political leaders' determination to maintain control over the state's internal trade. Furthermore, the swampy terrain south of Alexandria along the Potomac River made construction of a direct, north-south rail line difficult. Instead, most rail lines ran east-west and connected the foothills of the Blue Ridge Mountains with the tidewater region of eastern Virginia in order to transport natural resources from the interior to coastal port facilities accessible to ocean-going vessels.

Before the war, people traveling between Washington and Richmond had to take a steamboat or stagecoach fifty miles south to Aquia, Virginia, and then take a train an additional sixty miles south to Richmond. The only continuous rail route between the two capitals involved an indirect, 160-mile journey over two rail lines, the O&A and the Virginia Central Railroad. Over the next four years, Union and Confederate troops repeatedly fought battles along both railroads' rights-of-way for control of Manassas and other strategic rail junctions in the interior. U.S. Military Railroad personnel, oftentimes working within range of artillery and small arms fire, rebuilt rail lines and ran trains to support federal forces during many of these battles.

The O&A supported federal operations in northern Virginia in the early days of the war by transporting troops, supplies, and heavy equipment to encampments on the south bank of the Potomac. With the new rail connection across the Long Bridge, cars loaded in Baltimore moved directly to railheads in Alexandria. From there, locomotives moved them west along the O&A to supply dumps on the outskirts of Alexandria. The rail connection represented a vast improvement over the wagon trains that had supplied federal military forces in past wars.

In addition to repairing the rail line, Carnegie worked with Strouse and Sanford of the American Telegraph Company to repair the telegraph lines along the O&A.[29] By the First Battle of Bull Run on July 21, 1861, military telegraph lines extended approximately fifteen miles west along the O&A as far as Fairfax Station and Burke's Station, Virginia.[30] These lines provided nearly instantaneous communication between officers in the field and the War Department's telegraph office, and they markedly improved the ability of senior commanders to respond to the needs of subordinates. Additionally, the telegraph lines enabled Scott and Carnegie to monitor military railroad operations along the O&A and to resolve traffic snarls and other emergencies quickly.

The telegraph stations along the O&A provided Lincoln and his cabinet with their only source of information about the fate of federal forces engaged in the First Battle of Bull Run.[31] Lincoln's officials gathered in the War Department telegraph office to view dispatches from the battle. Late in the afternoon, they learned that federal troops were hastily retreating to Washington. Scott ordered his telegraph operators to remain in their stations despite the general retreat. Later, Charles Jacques, a sixteen-year-old operator stationed near Manassas, recalled, "Colonel Scott, in spite of my endeavors to close the office, still kept me there, telling me, if I left my post, he would have me shot."[32] Scott eventually permitted all operators and railroad personnel to retreat toward Alexandria in an orderly manner, preserving as much of their equipment as possible.

In the midst of the battle, Scott utilized the military telegraph to facilitate troop movements along the O&A. He had rail cars loaded with soldiers ready to depart for the battlefield and urgently telegraphed Brigadier General Irvin McDowell, informing him of their status:

> Do you want re-enforcements at Fairfax Courthouse? There are three regiments at Fairfax Station on the railroad, within three miles of you; and we have another regiment loaded on cars at Springfield Station, which can reach you in three hours, if you send them.
>
> We also have a regiment at the railroad station in Alexandria, which can reach Fairfax Court-House in four hours.
>
> Give instructions immediately.[33]

Unfortunately, McDowell did not reply to Scott's dispatch, and the Union defeat on the battlefield turned into a rout as panicked troops streamed back to the safety of the Potomac River. Scott, anticipating the growing threat to Washington by advancing Confederate troops, contacted Governor Andrew Curtin of Pennsylvania on his own authority and asked him to dispatch a force of twenty-five thousand Pennsylvania militia recruits. The men quickly mustered and traveled by train to the capital, bolstering the small number of troops defending the beleaguered city. Scott's managerial efficiency and quick thinking during the Bull Run debacle won him praise from the military establishment. Lieutenant General Winfield Scott, the commander of all Union forces, reportedly wished to give him a battlefield command. Scott declined the offer and continued serving as manager of the government railroad and telegraph network.[34]

Following Bull Run, Scott's duties increased dramatically, and he could not devote his full attention to railroad and telegraph affairs. Secretary Cameron had been lobbying Congress to create a new assistant secretary of war position. Congress finally approved his request on August 3, 1861. At a cabinet meeting, Cameron nominated Scott to fill the position. Lincoln enthusiastically responded,

"I am for Mr. Scott," and Scott received the appointment.[35] Others shared Lincoln's enthusiasm. Journalist Henry Jarvis Raymond declared, "Mr. Scott is gifted with a rare methodical energy that enables him to comprehend and execute with remarkable facility."[36] Though excited about the appointment, Scott had concerns about its impact on his role as the Pennsylvania Railroad's vice president. In an August letter to the PRR's board of managers, he indicated that he would resign his new position by the beginning of October 1861 and return to his old job if the directors desired. Board members expressed confidence in Scott's abilities and a month later, acting on a letter from Major General George B. McClellan, granted Scott an official leave of absence until the federal government no longer required his services.[37]

Casualties of War

Overwork and disease gradually reduced the cohort of Pennsylvania Railroad personnel that Scott brought into the War Department at the beginning of the conflict. Telegraph manager Strouse fell ill after having spent much of the summer of 1861 living in the field while overseeing line construction. Suffering from hemorrhaging in his lungs, he received a leave of absence and returned to Pennsylvania, where he died on November 17, 1861.[38] Carnegie also fell ill that summer while supervising railroad and telegraph operations in northern Virginia. Though his health gradually improved by the fall of 1861, Carnegie asked Scott for permission to return to his job as superintendent of the Pittsburgh Division of the PRR. Early fighting in the war had closed the B&O's trans-Appalachian trunk line through western Virginia, and PRR officials were struggling to accommodate much of the diverted traffic across their main line between Pittsburgh and Harrisburg. Carnegie believed that his managerial skills were needed to deal with the unprecedented volume of traffic and shore up the vital transportation corridor between the Midwest and the Atlantic Seaboard.[39]

As Strouse and Carnegie departed government service, new managers from outside the Pennsylvania Railroad establishment gradually assumed control over the federal government's railroad and telegraph networks. In the fall, Lincoln appointed McClellan commander of the Army of the Potomac, the main Union army in the East. In his summer campaign of 1861, McClellan had relied on a military telegraph network organized by Anson Stager, the general superintendent of the Western Union Telegraph Company, to outmaneuver and defeat Confederate military forces in remote northwestern Virginia (now West Virginia). McClellan was also familiar with railroad operations, having served as the Illinois Central's chief engineer and later vice president before the conflict.[40]

At the beginning of the war, the governors of Ohio, Indiana, and Illinois had seized control of all private telegraph lines running through the state and placed them under Stager's custody. The action was largely symbolic, as Stager already controlled one of the two major commercial lines running through the three states. Stager later took to the field with McClellan and established field telegraph offices to keep him in contact with military units scattered throughout the rough, mountainous terrain of northwestern Virginia. Scott had officially recognized Stager as the head of military telegraph operations in the Western Theater, but he had little official interaction with him during the summer of 1861.[41]

As commanding general, McClellan devoted particular attention to the new military telegraph network. By the fall, the government's network consisted of fifty small stations in northern Virginia, eastern Maryland, and the District of Columbia connected by over 280 miles of galvanized iron wire and serviced by eighty-three operators. Despite Strouse's heroic efforts during the spring and summer, the Military Telegraph Corps still lacked effective organization and leadership. Managers did not have military or governmental authority to issue orders and make financial decisions. Consequently, they continued to depend on commercial telegraph firms for supplies and technical assistance.[42]

Strouse's sudden departure also created a power vacuum within the government telegraph service. Throughout the East and Midwest, government telegraph agents working alone or with the aid of commercial firms competed against each other for supplies and equipment, and made contracts for line construction that varied wildly in cost from region to region.[43] As part of his all-encompassing efforts to reorganize Northern military forces, McClellan requested that the War Department create a centralized military telegraph bureau to oversee all telegraph activities across the country.

In response to McClellan's request, Scott asked Stager to come to Washington in September 1861 and submit an organizational plan for an independent military telegraph corps that could construct, operate, and maintain telegraph facilities in the field. In late October, Stager presented a plan that completely severed military telegraph and military railroad operations. Stager's plan called for "the appointment of a General Manager, whose duties shall be, under the advice and approval of the Secretary of War, to purchase, transport, and distribute all material required in constructing, maintaining, and operating Government telegraph lines."[44] Military telegraph managers would have authority to construct lines and negotiate with commercial firms for the use of their facilities when necessary. Finally, all members of the military telegraph service would be civilians and thus not subject to military authority or regulations.[45] Stager's plan represented a considerable departure from the integrated railroad-telegraph management framework originally put in place by Carnegie and Strouse.

Secretary of War Cameron and Assistant Secretary of War Scott both approved Stager's reorganization plan. The only objection came from Quartermaster General Montgomery C. Meigs, who argued that the head of the military telegraph service needed a military rank to provide him with the authority to requisition goods and equipment as necessary. Recognizing Meigs's wisdom in procurement matters, Scott arranged for Stager to receive appointment as a brigade quartermaster with the rank of captain. On November 25, 1861, the War Department officially appointed Stager "General Manager of the [Military] Telegraph lines."[46] He subsequently received promotion to Colonel of U.S. Volunteers for the United States Military Telegraph Corps.[47]

Scott had also been recruiting new managers for the military railroad network in the fall of 1861. After he lost Carnegie's services, Scott began looking for assistance from competent officials outside the close-knit PRR establishment. He appointed Thomas Canfield, former president of Vermont's Rutland & Washington Railroad, as manager of government railroad operations north and east of the Potomac River. For operations in northern Virginia, he selected Robert F. Morley, who left his position as president of the Allegheny Valley Railroad to work for the War Department. Scott tasked Morley with rebuilding and operating the Orange & Alexandria and Hampshire & Loudon Railroads, the two principal routes connecting Alexandria with points west and south.

Reporting to Secretary Cameron in November 1861, Morley stressed the unique challenges he faced in operating a military railroad. Unlike a civilian line, his had no "permanent sources of revenue to meet the current expenses and the decay of property." Much of his time involved repairing track and equipment damaged or destroyed by Confederate troops, rather than improving the efficiency of railroad operations.[48]

Morley's report highlights a fundamental problem that many civilian railroad and telegraph officials faced in the first year of the war. Managers attempted to operate military railroad and telegraph lines in Maryland and northern Virginia as if they were simply extensions of civilian lines, but military railroads and telegraphs were not businesses that could be managed by employees from behind desks. Instead, civilian officials had to take their prewar business management practices into the field of battle and adapt them to chaotic wartime operating conditions.

Reorganization

In the fall of 1861, Stager began reorganizing the Military Telegraph Corps to create a more efficient and valuable auxiliary for Northern military operations. At the same time, he distanced himself from the managers of the Northern military railroad network because he did not have firsthand experience in railroad operations. Stager's action began to unravel what had initially been a tightly integrated managerial framework based on PRR practices. For the rest of the war, military railroad officials faced the problem of operating rail lines without directly controlling the telegraph lines that ran alongside them.

Early in 1862, Cameron stepped down as secretary of war. Lincoln replaced him with Edwin M. Stanton, the former attorney general under James Buchanan. Stanton recognized Assistant Secretary Scott's administrative abilities and asked him to remain in government service. At the same time, the new secretary of war realized that the War Department needed additional staff to properly oversee its expanding wartime operations. Congress authorized two new assistant secretary of war positions and added forty-nine clerks to the department's payroll. Stanton selected a current War Department administrator for the first assistant secretary of war position. For the second, he picked his former law partner Peter H. Watson. Together, these two men assumed some of Scott's overwhelming workload.[49]

When he took office, Stanton solicited Scott's views regarding the administrative structure of the War Department, particularly the railroad and telegraph divisions. In a memorandum to Stanton, Scott strongly recommended improving administrative efficiency by reorganizing railroad and telegraph management within the War Department. In place of the newly independent Military Telegraph Corps and the existing government railroad network, Scott recommended a single unified transportation and telegraph department under the direct supervision of the secretary of war. The bureau would "provide for the construction or extension of such railroads or telegraphs as the wants of the military departments may require, and operate the same in such manner as may be necessary." A single bureau chief would be responsible for coordinating all railroad and telegraph operations, relieving the "Quartermaster-General's Department of the labor and responsibility of all matters pertaining to transportation by rail and water and telegraphic operations."[50]

Scott's reorganization plan sought to restore the coordinated railroad and telegraph managerial structure originally established by Carnegie and Strouse. Scott recognized that the Pennsylvania Railroad's prewar operating practices represented an effective model for coordinating railroad and telegraph management. Stanton accepted the general principles behind Scott's reorganization plan, but he did not follow through with any of Scott's specific recommendations. Had Scott's January 1862 plan been adopted in full, the War Department would have continued to oversee an efficient and highly integrated railroad and telegraph network.

Stanton did succeed in obtaining congressional recognition for the United States Military Railroad network. On the last day of January 1862, Congress enacted a statute authorizing the president to seize control of any railroad line within the United States to regulate the firm's operations and impose military authority on its personnel. These provisions provided the federal government with legal authority to compel civilian railroads to cooperate with the War Department. Additionally, the act provided statutory authority for the War Department to operate seized southern railroads as part of the USMRR network.[51]

Stanton used his new congressional authority to create a separate, largely autonomous railroad transportation division within the War Department. While Scott still exercised nominal authority over all railroad matters, he no longer directed day-to-day military railroad operations. Instead, Stanton once again looked beyond the close-knit Pennsylvania Railroad fraternity and selected Daniel C. McCallum as the new "military director and superintendent of railroads" on February 4, 1862.[52]

McCallum seemed like an ideal choice as USMRR director. The innovative and flexible telegraphic management practices that he had developed as general superintendent of the New York & Erie Railroad in the mid-1850s seemed well suited to the chaotic operating conditions on the USMRR lines in the East. His pioneering management model had allowed civilian officials to receive up-to-the-minute information on train locations and to coordinate train movements across the entire Erie network. These capabilities offered clear benefits to military railroad managers as well.[53]

McCallum soon discovered that railroad operations in war-torn northern Virginia bore little similarity to peacetime operations in upstate New York. He found that he had no control over the U.S. Military Telegraph network as the director of the USMRR. He could not build telegraph lines along railroad routes in Virginia, nor could he use existing military telegraph circuits solely for railroad business. In a letter to Quartermaster General Meigs on April 7, 1862, McCallum listed a telegraph line wholly controlled by U.S. Military Railroad managers as one of the critical requirements for supporting military railroad operations in the field: "From the unavoidable irregularity of demands made upon us for transportation, trains cannot usually be run by time schedule. The use of the telegraph is indispensable for that purpose. This will require the construction of an independent wire."[54] Unfortunately for McCallum, control over all military telegraph wires remained firmly in the hands of Stager and the Military Telegraph Corps throughout the war.

McCallum soon faced criticisms for managing military railroad activities from his Washington headquarters, instead of directing operations in the field. On several occasions in the spring of 1862, Major General Irvin McDowell, commander of the Department of the Rappahannock, complained to Stanton about deteriorating operating conditions on the Orange & Alexandria. McDowell also criticized McCallum's lack of progress rebuilding the Fredericksburg Railroad. He noted: "The service would be benefited could Colonel McCallum be out here. There is no way of producing harmony of action between the command and the railroad department."[55] McCallum acknowledged that problems existed but asked for a few days to remedy the situation. Two days later, after receiving reports from field commanders about supply shortages, Stanton pressed McCallum to complete work on the O&A.[56] On May 11, Major General McDowell bluntly informed the War Department, "Colonel McCallum does not take very good care of us."[57]

McDowell depended on the military railroad lines of northern Virginia for many of his supplies. With the Fredericksburg Railroad inoperable, and the O&A barely functioning, he could not advance beyond his supply bases. Meanwhile, McClellan's massive Army of the Potomac had landed on the Yorktown Peninsula and was marching inland to take Richmond from the east. McDowell's smaller Army of the Rappahannock was supposed to march south toward Fredericksburg, Virginia, to take pressure off McClellan's western flanks. Without functioning railroads, however, McDowell could do little to assist McClellan.

By late April, Stanton concluded that McCallum was ill suited to the task at hand. To alleviate the growing transportation crisis, Stanton sought out the services of Herman Haupt, former general superintendent and chief engineer of the Pennsylvania Railroad during the early 1850s. By 1862, Haupt was overseeing the excavation of the 4.75-mile-long Hoosac Tunnel in western Massachusetts, the longest railway tunnel in North America at the time. Stanton urgently telegraphed Haupt on April 22, 1862, and asked him to come to Virginia to take charge of all railroad operations in the Department of the Rappahannock. Haupt accepted the offer but cautioned the Secretary of War: "I have no military or political aspirations, and am particularly averse to wearing the uniform; would prefer to perform the duties required without military rank if possible, but if rank is essential as a means to aid in the performance of duty, I must acquiesce. . . . If I take the position you have so kindly offered, it will be with the understanding that I can retire whenever, in my opinion, my services can be dispensed with."[58] Haupt immediately departed for the front lines and arrived at Aquia Creek, the terminus of the Fredericksburg Railroad on the western bank of the Potomac River, a few days later.

After assessing the situation and meeting with McDowell, Haupt traveled north to Washington to confer with Stanton. On April 27, the secretary of war appointed him chief of construction and transportation of the Department of the Rappahannock with complete authority over all railroad activities within the department's jurisdiction. The following day, Stanton issued orders giving Haupt absolute operational autonomy within the War Department. Haupt answered directly to Stanton, not McCallum.[59]

Stanton's actions disrupted the chain of command within the U.S. Military Railroad department. McCallum and Haupt both held the rank of colonel, but railroad transportation within the Department of the Rappahannock technically fell under McCallum's purview as military railroad superintendent. Fortunately, both managers found a way to resolve the awkward situation. In his memoirs, Haupt described the compromise solution: "McCallum was a splendid office man, thoroughly familiar with every detail of requisitions, accounts and red tape, which I was not. Moreover, I did not care to learn, so it did not take long to come to a perfect understanding and division of duties which suited us both. McCallum took the office and I took the field."[60] The informal arrangement continued for over a year until Haupt officially left military service in the fall of 1863.

Whereas McCallum had struggled to transition from civilian to military railroading, Haupt thrived under the chaotic wartime operating conditions, possibly because he spent much of his early career overseeing field operations in central and western Pennsylvania for the PRR. At Aquia Creek, he quickly set to work reconstructing the rail line to Fredericksburg. With a construction corps consisting largely of untrained soldiers, civilians, and former slaves, he rebuilt two major bridges destroyed by retreating Confederates. The first bridge, 150 feet in length, required a mere fifteen hours to complete, while the 400-foot-long, four-story Potomac Run Bridge required nine days and over two million board feet of timber. Despite limited resources, Haupt finished the railroad to Fredericksburg by May 19, twelve days after he arrived on-site.[61]

President Lincoln visited the site of the Potomac Run Bridge shortly after Haupt's workers finished construction. After seeing the structure, the President quipped: "I have seen the most remarkable structure that human eyes ever rested upon. That man, Haupt, has built a bridge across Potomac Creek . . . over which loaded trains are running every hour, and upon my word . . . there is nothing in it but bean-poles and cornstalks."[62]

Confederate advances in the Shenandoah Valley and McClellan's failure to capture Richmond during the Peninsula Campaign forced McDowell to abandon his drive to Fredericksburg and to retreat north to protect Washington. In late May, he gave Haupt the task of reorganizing and repairing the railroad lines connecting Alexandria with Manassas Junction and the Shenandoah Valley. Once again, Haupt dealt successfully with numerous operating problems on these lines.

Haupt approached railroad operational management very differently than had McCallum. Unlike the former Erie superintendent, Haupt refused to rely on the Military Telegraph Corps lines for directing train movements along the war-torn rail lines of northern Virginia. He had good reasons for avoiding the telegraph. After the war, Haupt recalled, "When the [telegraph] wires were not down they were in use by the military authorities, who would not allow them to be interrupted." He harshly criticized McCallum for continuing to depend on the independent military telegraph during the early years of the war: "Colonel [Daniel] C. McCallum . . . was a thorough master of transportation, when he could be relieved from military interference and have a telegraph line under his entire control—conditions that did not exist previous to my charge."[63] Haupt also insisted on managing railroad operations from the front lines, not from an office in the War Department. In all, he believed that his rudimentary managerial practices were more flexible and better suited for wartime conditions than McCallum's sophisticated telegraphic management model.

In a September 1862 letter to Major General Henry Halleck, commander in chief of federal military forces, Haupt presented a critique of the Northern military railroad network and offered a proposal for reforming its operations. He drew Halleck's attention to the inadequacy of the telegraph as a tool for overseeing military railroad operations. He noted, "[Telegraphic dispatching] invariably leads to difficulty, and in case of any derangement to the delicate mechanism of the telegraph, puts an end to all business and blocks everything upon the road." While Haupt conceded that the telegraph could prove "valuable as an auxiliary, [it] should not be used as a principal."[64] Instead, Haupt advocated strict adherence to operating schedules on all military railroad lines. Deviations from the schedule should not be permitted. He concluded his letter to Halleck by admonishing the major general, "It is desirable that uniformity should be introduced in the management of all railroads used for military purposes." Halleck later issued strict orders granting Haupt supreme authority over all military railroad operations and barring military officials from interfering in any way with train movements.[65]

In practice, Haupt instituted a strict, rule-based operating model. He ordered that he, his deputy, or an authorized train dispatcher must give all train orders personally. Military officers could not interfere with train movements or countermand standing orders. All trains would run on set schedules with the telegraph being used only for administrative business, not for overseeing train movements. If trains became delayed or interfered with other trains' scheduled running times, Haupt dispatched trusted railroad personnel on horseback to issue necessary orders and resolve the problem. Delayed train crews were under standing orders to proceed slowly down the line after sending a flagman ahead of their train to scout for oncoming traffic. In cases of extreme urgency, Haupt utilized convoys for moving trains rapidly from one location to the next over a single track. By sending all railroad traffic in one direction over the course of a few hours, he avoided complicated scheduling problems and maximized traffic flow.[66]

During his time in the military railroad service, Haupt insisted that employees adhere strictly to his directives. In a December 1862 letter to military railroad personnel, Haupt reminded the men, "The military railroads have utterly failed to furnish transportation to even one-fifth of their capacity when managed without a strict conformity to schedule and established rules." Officers should not be allowed to interfere with scheduled train operations, and a "clock should be conspicuously placed at each station" to indicate starting times for trains. Only through obedience to these directives could military railroads function like their civilian counterparts. Haupt summed up his managerial prerogative boldly, "Punctuality and discipline are even more important in the operations of a railroad than to the movements of an army."[67]

While Haupt and McCallum both emphasized the need for order and discipline in their military guidelines, they were clearly committed to incompatible managerial models. For Haupt, only strict adherence to operating rules and personal oversight of frontline activities could ensure "punctuality and discipline" in railroad operations. McCallum, on the other hand, continued to utilize telegraphy to administer military railroad operations, despite the military railroad's lack of control over telegraph facilities along its route.

In December 1863, Stanton asked McCallum to review military railroad operations in Tennessee. Up to that point in the war, military railroads in the Department of the Cumberland had operated largely independently of those in northern Virginia. McCallum had neither devised the operating practices employed by western railroad managers nor provided direct supervision to these officials.[68] Following his review, McCallum issued reports condemning military railroad management in Tennessee as "decidedly defective."[69] He found railroad personnel acting "with neither system, order, nor discipline."[70] McCallum placed the blame on incompetent military officers and railroad managers with a "seeming want of ability to comprehend the magnitude of the undertaking." For McCallum, the solution to the managerial problems lay in thoroughly reforming military railroad organization in Tennessee. He proposed an administrative structure similar to the operational framework that he had imposed on the Erie Railroad in the mid-1850s. The lines would be divided into multiple divisions. Divisional managers would periodically send reports on railroad conditions and train movements directly to the head of the operations department via telegraph. McCallum concluded that such a model would "insure perfect discipline and full co-operation throughout."[71]

While McCallum's proposed operating plan for the Department of the Cumberland reflected a deep understanding of operational management practices, it depended on free and unfettered access to telegraph facilities along rail lines. As had been the case in northern Virginia, McCallum's plan proved untenable, since all telegraph facilities in Tennessee remained under the control of the U.S. Military Telegraph Corps. Once again, the independence of both agencies interfered with McCallum's proposed reforms.

The surviving portions of a daily journal of events maintained at the U.S. Military Railroad depot in Alexandria in late 1863 highlight the challenges that McCallum faced as he continued to depend on telegraphic dispatching for USMRR operations. On November 19, the log keeper, most likely a train dispatcher, discussed track conditions between Alexandria and Brandy Station, Virginia, on the O&A and listed where trains could, and could not, use the daily schedule. At the end of the entry he noted, "Trains wait for [telegraph] orders until military business gets through," meaning that military dispatches had tied up the telegraph lines, and the train dispatcher could not contact station agents with starting times and meeting points for trains on the rail line.[72] The next day, the dispatcher praised the schedule for keeping the line "straight and running today," since the "[telegraph] wire this morning was in bad order and worked very hard all day."[73] A few weeks later, the writer complained of "an uncomfortable day with the works and matter running roughly." He described a rear-end collision caused by one train following another too closely. While the damage was slight, "the delay has thrown every other train off time."[74] Without the military telegraph network under his control, McCallum and his subordinates could do little to resolve such daily problems. Instead, they were forced to depend on Haupt's schedule-based management practices.

McCallum clearly recognized the difficulties caused by the administrative division between the federal government's railroad and telegraph departments. He complained about the situation in his annual report for the 1864 fiscal year:

> Great embarrassment and often serious delays occurred from lack of proper telegraph facilities, resulting from exclusive control of the military telegraph by another department. . . . The telegraph is to a railroad what the nerves are to a human body, and the condition of a person whose nerves are controlled by the will of another may be readily imagined. . . . When nine-tenths of the work done by the telegraph is railroad business and often its promptness and correctness are of vital consequence, it would seem proper for the railroad managers to control the men whose services are so important to them.[75]

In McCallum's view, Stanton's policies had effectively negated his ability to exercise managerial control over subordinates in the field. McCallum's train crews had to wait for hours because military telegraph traffic took priority over train dispatching orders. He could not deal with unforeseen problems, such as wrecks, as he had been accustomed to doing during his civilian career. Without free access to the telegraph, McCallum had no tangible managerial power. On the contrary, he felt as though forces beyond his control were manipulating his railroad department.

McCallum's 1864 annual report clearly indicates that he found military railroad management without the benefit of telegraphy both frustrating and inefficient. Despite Scott's and Carnegie's efforts to maintain a tight link between military railroad and telegraph operations at the beginning of the war, General Manager Stager had effectively distanced military telegraph management from railroad operations. Telegraph personnel continued to work along military railroad routes and provide communication services for railroad operations, but McCallum had to appeal to Stager every time he needed additional operators to man new stations or to provide relief for overworked dispatchers at busy points on the railroad lines.[76] McCallum firmly believed that this arrangement had to be radically overhauled to address its shortcomings.

Consolidating Management

In conjunction with his efforts to reform military railroad operations in the field, Chief of Construction Haupt (like McCallum) wanted to conduct a broader administrative reorganization of the far-flung United States Military Railroad network. By 1863, the War Department nominally controlled military railroad lines scattered across much of the country. In addition to the core lines in northern Virginia, the federal government operated lines in the Gulf region and the Department of the Cumberland. These outlying railroad networks functioned largely autonomously. Quartermaster Department officials managed some lines, while Army staff officers administered others.

Haupt pushed Secretary of War Stanton to remove the USMRR from the Quartermaster Department's control and to place it in a new bureau within the War Department. The railroad bureau would consolidate all military railroad affairs under a single administrator who would report directly to the secretary of war. Haupt believed that the new bureau would give civilian managers greater freedom from US Army bureaucracy and would allow them to operate their military rail lines as they saw fit.[77]

Unlike the War Department railroad bureau that Scott had proposed in 1862, Haupt's bureau did not include military telegraph operations. This omission further highlights Haupt's general disinterest in telegraphy as a tool for monitoring and controlling train movements. While McCallum saw an urgent need for a unified telegraph and railroad department in the War Department, Haupt did not believe that it was necessary for railroad and telegraph officials to coordinate their actions. His old-school management practices had worked over the past year, and he had no reason to believe that they would not continue to work for the remainder of the war.[78]

In mid-1863, Haupt presented his reorganization plan to Secretary Stanton and offered to serve as the new head of the railroad bureau, but his star was quickly fading in the War Department. Since he had entered military service in 1862, Haupt had been dividing his time between serving as construction chief for the USMRR and overseeing his financial interests in the Hoosac Tunnel construction project in Massachusetts. Massachusetts governor John Andrew disliked Haupt and had spent the past two years trying to convince the state legislature to remove him as head of the project. Andrew was a political ally of Stanton, and the secretary of war agreed to pressure Haupt to end his involvement in the tunnel project. On September 1, 1863, Stanton wrote to Haupt giving him five days to formally accept a full-time military commission that had been offered to him or to leave federal service. If Haupt accepted the commission, he would be disqualified from working on the Hoosac project for the duration of the war. Haupt refused to give up work on the tunnel project, and Stanton relieved him of his command a few days later.[79]

After Haupt's departure from military service, McCallum assumed many of his responsibilities. McCallum soon consolidated his power as the head of military railroad operations in the East. Following an inspection of the Department of the Cumberland in the winter of 1863–64, Stanton appointed McCallum general manager of all military railroads in the Departments of the Cumberland, Tennessee, Ohio, and Arkansas.[80] While he continued to exercise control over eastern military railroad operations, McCallum's new task involved reforming railroad management in the Western Theater of combat. McCallum quickly expanded the size of the Construction Corps, originally created by Haupt, and put them to work rebuilding and expanding the rail lines supporting federal advances in Tennessee, Mississippi, and Georgia.[81] He eventually removed numerous ineffective local railroad managers and replaced them with eastern managers well versed in military railroad operations.

After the Army of the Potomac's advance into central Virginia in the spring of 1864 under Major General George Meade, railroad lines in northern Virginia became less critical for supporting federal troops in the field. Coordinating with the Army as it marched south, USMRR personnel put new rail lines into operation closer to the front lines. The Construction Corps repaired the City Point & Petersburg and the Richmond & Petersburg Railroads along the James River to provide supplies for military forces besieging Richmond and Petersburg in 1864–65. At the same time, the USMRR abandoned the Fredericksburg Railroad from Aquia Creek on the Potomac, as it no longer directly served military needs.[82]

In July 1864, Congress approved a plan to reorganize the Quartermaster Department into nine separate divisions. One of these sections included all military railroad and telegraph operations. While the reorganization placed both McCallum's Military Railroad department and Stager's Military Telegraph Corps under Quartermaster General Meigs's administrative command, the change had minimal impact on railroad and telegraph operations. Both groups continued to maintain distinct identities and administrative structures. McCallum did not gain control over the military telegraph network. For the remainder of the war, he continued to compete with military officials for access to limited telegraph resources.[83]

Civilian Railroads and Wartime Operations

Civilian railroads in the East and Midwest handled unprecedented volumes of passenger and freight traffic during the Civil War. The federal government contracted with northern railroads to move men, animals, and supplies across the country in support of its war efforts. The Pennsylvania Railroad transported nearly a million troops for the federal government between 1861 and 1865, in addition to approximately four million civilian passengers. The Illinois Central (IC) handled over half a million soldiers during the same period. Similarly, the volume of military and civilian freight transported on the PRR doubled from 1,500,000 tons in 1861 to nearly 3,000,000 tons by 1865.[84]

Wartime freight and troop traffic soon stressed northern railroad networks to the breaking point. In the East, major trunk lines, such as the New York & Erie Railroad and the PRR, bore the brunt of the increase, but smaller companies faced similar problems. PRR affiliates, such as the Northern Central Railway and the Philadelphia, Wilmington & Baltimore Railroad (PW&B), rapidly expanded their operations to handle the traffic fed to them by the PRR. The PW&B expanded its locomotive pool by 25 percent during the war and added 230 freight cars to its prewar fleet of 674 cars to cope with government and military traffic.[85]

Midwestern railroads faced similar burdens. The Pittsburgh, Fort Wayne & Chicago (PFW&C) experienced nearly a 100 percent increase in passenger volume during the war as it moved troops from Chicago to mustering points in the East. The Illinois Central, the Michigan Central, and the Chicago, Burlington & Quincy (CB&Q) also faced overwhelming levels of freight and passenger traffic. The Illinois Central moved so many troops and supplies from Chicago to military camps near Cairo, Illinois, that it could not maintain a pool of cars to handle routine freight and passenger demands.[86]

Managers quickly discovered that peacetime operating practices could not alleviate wartime traffic burdens. Dramatic increases in military passenger and freight traffic, usually before and after major battles, often overwhelmed fixed schedules and led to congested rail lines and train accidents. The number of passenger train accidents rose precipitously during the war. In 1861, 63 accidents injured 459 passengers and crew and killed 404. By 1864, the number of accidents had grown to 140. These accidents accounted for 1,846 injuries and 404 deaths.[87]

Increasing traffic congestion forced managers to look for safer and more flexible operating practices. Officials on a few eastern railroads that already employed daytime telegraphic dispatching began using it around the clock. Others reluctantly adopted it for the first time, but senior managers on some of the leading railroads in the East and Midwest dragged their feet. Many were concerned about the added costs of employing telegraphy for railroad operations.

Following his service with the USMRR, Carnegie returned to his position as superintendent of the Pittsburgh Division of the PRR. He soon realized that the PRR's main line between Pittsburgh and Harrisburg had become dangerously congested. In his autobiography, Carnegie recalled that he decided to employ a night dispatcher to keep traffic flowing smoothly twenty-four hours a day. His actions had a broader financial impact than hiring a single employee. The PRR would be forced to keep telegraph offices open throughout the night at many smaller stations along the line to provide the dispatcher with current information on train movements. Given these escalating costs, it is hardly surprising that senior PRR officials did not enthusiastically support Carnegie's unauthorized actions.[88]

Managers on other eastern lines were no more enthusiastic about spending money on continuous dispatching. In late July of 1863, as the federal government tried to reinforce and resupply military forces in the East following the Battle of Gettysburg, Gustavus A. Nicolls, the general superintendent of the Reading Railroad, wrote to John O. Stearns, superintendent of the Central Railroad of New Jersey, complaining about the Jersey Central's failure to employ night operators on its busy main line between Jersey City and Easton, Pennsylvania. Reading trains exchanged passenger and freight cars with Jersey Central trains at Easton. Without twenty-four-hour telegraph service, Reading officials had no way to notify Jersey Central managers to prepare for unscheduled evening traffic. This caused serious delays and blocked the Reading's line at Easton. Nicolls begged Stearns to keep the Central's telegraph offices open later than 8 p.m. when both lines were inundated with military traffic, but Jersey Central officials ignored his request.[89]

Over a year later, Nicolls again wrote to Stearns about the importance of night dispatchers: "Allow me to press upon you the necessity of having night telegraph operators on your line between Jersey City and Easton—now that we have important express trains to put through safely and promptly." He reminded Stearns that the Reading had employed night dispatchers on its busy Lebanon Valley Railroad between Harrisburg and Philadelphia "for years" with "an excellent result." Nicolls could not understand why Jersey Central officials would resist expanding their dispatching operations to cope with wartime traffic.[90] He failed to realize that Jersey Central officials, like senior PRR managers, likely did not want to pay the additional costs that twenty-four-hour service would entail.

Wartime congestion forced officials on some newer railroads to adopt telegraphic dispatching for the first time. An independent commercial telegraph line had paralleled the Illinois Central since its completion in 1855, but railroad officials had viewed telegraphic dispatching as a "dangerous expedient." They had used it on only a few occasions prior to the war.[91] In 1863, as the IC's main line grew increasingly congested, managers began using the commercial line to send train orders. The firm's growing dependence on telegraphic dispatching overwhelmed the line. In 1863, IC president William Osborn requested that John D. Caton, president of the Illinois & Mississippi Telegraph Company, erect a second telegraph wire exclusively for railroad use, since the commercial telegraph wire was "fully engaged by the Government and the public."[92] Later that year, a winter storm damaged part of the telegraph line. Osborn wrote to Caton about the problem, noting, "Now that our trains are moved by Telegraph, this disarrangement causes the most serious consequences."[93] As officials feared, telegraphic dispatching forced them to rely on outside parties to keep their railroad running smoothly and introduced unforeseen managerial challenges.

Officials on other lines struggled with the broader impact of telegraphic dispatching as well. At the beginning of the war, CB&Q general manager Robert Harris was reluctant to introduce telegraphic dispatching. He believed that the CB&Q could accommodate wartime traffic through traditional operating practices. The rapidly expanding volume of freight and passenger traffic on the company's rail lines, however, left him with no alternatives.[94] In 1863, Harris entered into negotiations with Caton's telegraph company for a second telegraph line for the "exclusive use, management, and control of the R.R. Co."[95] The new line allowed the CB&Q to implement round-the-clock telegraphic dispatching services to meet wartime demands.

After Harris introduced dispatching on the CB&Q, his subordinates sought to overhaul the company's operational management practices to address new risks posed by telegraphy. The railroad's superintendent of the telegraph, Frederick H. Tubbs, instituted a telegraphic train order arrangement that supplemented the CB&Q's basic operating rules. This mixture of older and newer practices later became commonly known as the "American system" of train dispatching.[96] It called for a single dispatcher to monitor traffic on each division. When delays or increased traffic disrupted the daily timetable, the dispatcher issued new train movement orders to telegraph operators. Operators passed these new orders to the train crews who were affected by the delays. One industry observer praised the advantages of the new operating model: "By placing the movement of trains under the complete and absolute control of one man the chances of accident are reduced to the lowest minimum."[97] He failed to note that the management model lacked a mechanism for preventing accidents if the dispatcher made an error. Despite this flaw, Tubbs continued to improve his version of the American system in the postwar era until many contemporaries held it up as a model for efficient telegraphic dispatching.

While the American system offered an efficient way to monitor train movements and prevent traffic jams, it introduced many unforeseen operating risks.

"Chief among these [risks]," writes Robert B. Shaw, "was the establishment of a chain of human intermediaries through whom train orders were transmitted. . . . This process multiplied the scope for error at the same time it broadened responsibility."[98] Station agents and telegraph operators had little influence over train movements under the old managerial framework, but under the American system they could cause accidents by failing to properly communicate train orders to crews. Likewise, dispatchers could inadvertently send trains crashing into each other if they issued conflicting orders. While superintendents ultimately bore responsibility for accidents on their divisions, they exercised little direct control over many of the employees (both competent and incompetent) responsible for routine train movements under the new "dispatching chain."

A major wreck on the Erie Railroad highlighted the dangers posed by intermediaries along the dispatching chain. On July 15, 1864, a transport train carrying Confederate prisoners of war collided with a freight train near Shohola, New York. The prisoner train had departed the docks in Jersey City later than expected. Further west, a telegraph operator at Lackawaxen, New York, had failed to pass along train orders to the conductor of a local coal train informing him of the delay. Instead, the operator allowed the coal train to enter the main line from a sidetrack. A few miles down the line, the transport train collided with the coal train. In the ensuing wreck, fifty-nine prisoners died, along with nineteen guards. Such accidents raised serious questions about whether the advantages of the American system outweighed the risks.[99]

At the end of the war, managers throughout the East and Midwest evaluated the economic and organizational costs and benefits of telegraphic dispatching, and many concluded that it was necessary for coping with increased traffic on their lines. Few officials, however, had the expertise or necessary financial resources to correct the inherent flaws in the new management model. Consequently, many officials continued to run trains under poorly developed wartime operating protocols instead of implementing broader reforms in the postwar era.[100]

Conclusion

The Civil War's impact was felt broadly both within military and civilian realms. By war's end in April 1865, the United States Military Railroad had expanded into a network encompassing nearly twenty-one hundred miles of track and employing roughly twenty-five thousand civilian employees.[101] When it ceased operating in early 1866, it was the single largest railroad organization in the world. Ten years would pass before the Pennsylvania Railroad overtook it in length of track and number of personnel.[102] Over the course of the war, the Military Telegraph Corps controlled over fifteen thousand miles of telegraph wire and employed roughly one thousand operators.[103] Both networks provided valuable communication and logistical support for federal forces fighting a war that stretched halfway across a continent, as well as a training ground for thousands of young men who would go on to seek jobs in the civilian world at war's end. Military telegraph historian William R. Plum estimated that the Military Telegraph Corps sent over 6,500,000 telegrams during its four-year existence.[104] Anson Stager perhaps provided the most succinct characterization of both organizations in his 1865 annual report: "The military railroads and the military telegraph have been great auxiliaries to the gigantic and successful efforts of the Government in suppressing the rebellion."[105]

The Civil War revealed US Army officers' lack of knowledge and training in the fields of railroading and telegraphy. From the very beginning of the conflict, civilians with backgrounds in railroading ran the federal government's rail lines and oversaw a rapidly expanding military-industrial complex. Thomas A. Scott made the reality of the situation clear to Secretary Stanton early in 1862 when he argued that few military officers had any useful knowledge of railroad practices and that government transportation should be left in the hands of civilians.[106] Four years later, Daniel C. McCallum echoed Scott's argument in his summary report of USMRR operations during the conflict. After detailing the unique difficulties faced by military railroad personnel operating in the field, McCallum stated, "Management of railroads is just as much a distinct profession as is that of the art of war, and should be so regarded."[107]

The war highlighted the conflicting opinions of civilian railroad authorities regarding the best practices for managing railroad operations. Herman Haupt's inherent distrust of telegraphy for train dispatching strongly influenced his managerial decisions while in federal service. Like many antebellum railroad officials, Haupt viewed telegraphy as unreliable and ill suited for operational management purposes. After the war, a few civilian railroad officials followed Haupt's wartime example and continued to shun telegraphy in favor of timetables and strict running rules.

While McCallum's telegraphic management practices proved unsuccessful in the early years of the conflict, they eventually demonstrated their worth by the end of the war. Stager's final report on the Military Telegraph Corps justified McCallum's persistent advocacy of telegraphy for railroad operational management. In the report, Stager remarked matter-of-factly: "The military telegraph has been an invaluable assistant in the construction and operation of the various military railroads. Trains have been run and many of the [rail]roads operated almost exclusively by telegraph."[108] Stager's comments reveal that McCallum did succeed in using telegraphy for train dispatching on some military rail lines. While he never gained control of the military telegraph network, he did put it to effective use at times. McCallum's views on railroad management most certainly had an impact on military railroad operations, and his cohort of fellow managers returned to the civilian railroad industry inspired by the potential of telegraphy for improving operational safety and efficiency.

The war marked a point of transition for managers in the American railroad industry. In the postwar era, increasing numbers of railroad officials accepted telegraphy as a powerful tool for waging war on the tyrannies of time and space in daily railroad operations. Yet, the war did not lead to universal acceptance of the device within the civilian railroad community. Its gradual adoption by railroad managers in the decades after the war remained contingent on a variety of economic, technical, and political factors.

The American System

> The time is fast approaching when railroads will be required . . . to establish and maintain an efficient telegraph system for the safety of their patrons, even more than for the facility afforded for working the [rail]roads systematically and effectively.
>
> Editor, *The Telegrapher*, 1872

Daniel C. McCallum's tenure as head of the United States Military Railroad revealed to observers across the country that the telegraph could be a critical tool for managing railroad operations. Though it was far from perfect, he and his subordinates depended on it to dispatch trains under chaotic operating conditions. Railroad officials like McCallum's rival, Herman Haupt, who refused to employ the telegraph, increasingly appeared to be out of touch with trends within the civilian railroad industry as a whole during the 1860s. Like the military lines, many civilian railroads in the North and Midwest were overwhelmed with freight and passenger traffic throughout the conflict. A few civilian officials, echoing Haupt, continued to depend on antebellum time- and rule-based operating practices to keep traffic running. Others reluctantly expanded prewar, part-time telegraphic dispatching practices to accommodate round-the-clock service. The majority of officials hastily adopted telegraphic dispatching on their lines for the first time as wartime traffic exceeded the capacity of their networks.

Few managers radically overhauled their operating practices when they adopted telegraphic dispatching during the war. Instead, most simply appended basic dispatching rules to their prewar employee rule books, thus creating what came to be known as the "American system" of train dispatching. The new operating framework enabled managers to move trains more expeditiously, but it did not incorporate physical or organizational safeguards to prevent collisions. Instead, it placed operational safety in the hands of a cadre of overworked and underpaid train dispatchers and telegraph operators, critical intermediaries in the "dispatching chain." Despite the risks, the new management practices alleviated much of the wartime traffic congestion and cost far less than adding additional railroad tracks.[1] It proved to be so successful that many officials continued to depend on it after the war.

Managers were not blind to the risks posed by the American system. Serious railroad accidents throughout the 1860s and early 1870s highlighted its

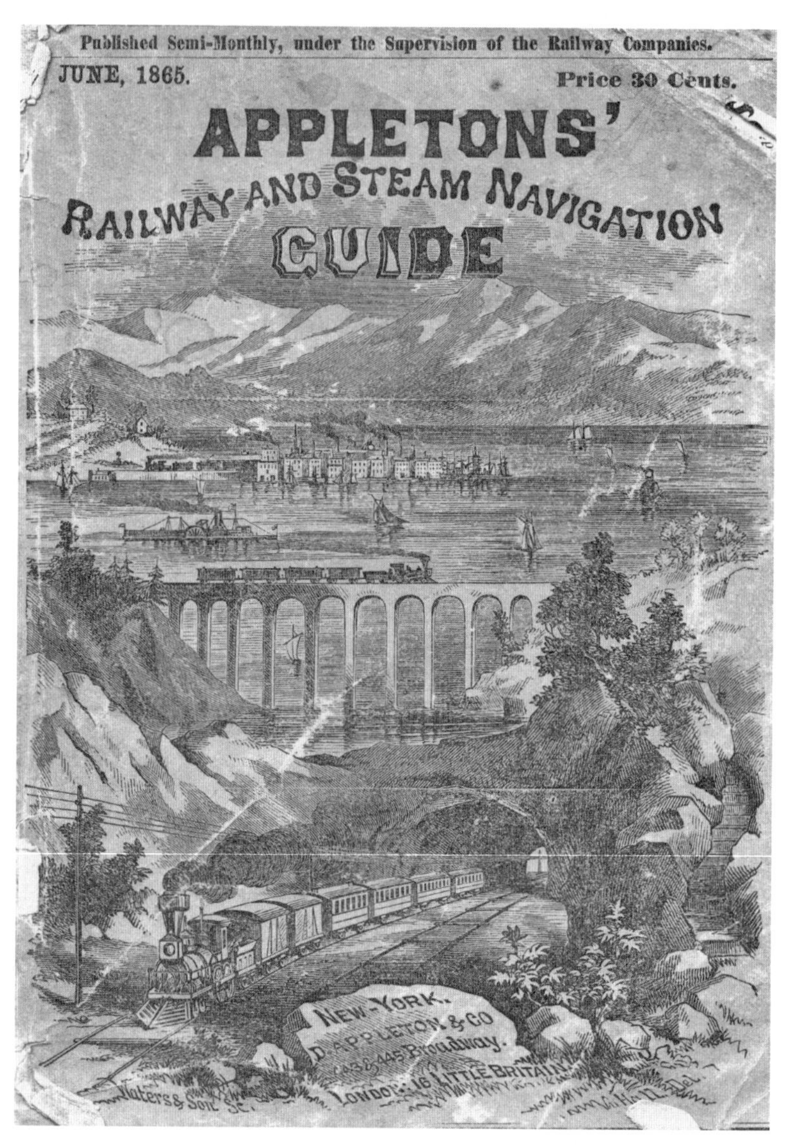

By the end of the Civil War, telegraph lines had become a common sight along American railroad rights-of-way, as the cover of the April 1865 edition of *Appleton's Railway and Steam Navigation Guide* illustrates. Many railroads had adopted telegraphic train dispatching during the war, but they depended on Western Union and other commercial telegraph firms of the era for telegraph service. Consequently, railroad managers devised operating practices that employed the telegraph in a limited manner. The so-called American system increased the efficiency of railroad lines but did little to ensure safe operating conditions. *The Schwantes-Greever-Nolan Travel and Transportation Ephemera Collection, St. Louis, MO*

shortcomings, particularly when compared to the more sophisticated telegraph-based block signaling and dispatching practices that were becoming increasingly common in Britain during this era.[2] A few well-capitalized American companies began experimenting with telegraph-based block signaling and other telegraphic train management practices during these years. Officials on poorly capitalized railroads, however, could not afford to implement costly operational reforms in the aftermath of the Panic of 1873. Instead, many came to depend on the inexpensive, and often poor quality, telegraph service that they received primarily from the Western Union Telegraph Company to operate their rail lines.

Difficult economic circumstances limited American railroad officials' reform options during the 1870s. Despite increasingly vocal demands for change from public officials and industry reformers, railroad officials had little choice but to loudly defend the expedient and inexpensive American system of telegraphic train management.[3]

Managerial Intransigence in Postwar New England

After the war, a few railroads, primarily in the Northeast, continued to stand firm in their opposition to telegraphic train dispatching. New England managers were not forced to adopt telegraphy during the war, owing to a number of unique managerial and economic factors. First, they had pioneered effective train operating practices in the pretelegraph era, and they did not believe that telegraphy offered significant safety or efficiency advantages. Second, officials had already faced the challenge of dramatic traffic increases before the war, and they had resolved the problem by double-tracking busy sections of their relatively short rail lines. Finally, New England railroads did not experience overwhelming levels of wartime traffic and continued to operate trains according to fixed schedules.[4]

New Englanders made no secret about their distrust of telegraphy during the Civil War era. "No Trains Run by Telegraph," the Stonington Line of Connecticut proudly advertised in a mid-1860s placard, along with "Reclining Chair Cars on All Express Trains."[5] The Stonington Line's advertisement clearly illustrated company managers' hostility to telegraphy. In addition, the advertisement implied that officials believed that New England passengers had anxieties about traveling on railroads that used telegraphic dispatching and would be more likely to patronize a company that employed traditional operating practices.

Many New England officials remained uncertain about telegraphic dispatching despite a major postwar passenger train accident that drew public attention to the region's antiquated operating practices. In late August of 1871, heavy passenger traffic inundated the Eastern Railroad of Massachusetts, a busy commuter route between Boston and Portland, Maine. End-of-summer travelers had ballooned the number of passengers utilizing the rail line from an average of 110,000 per week to well over 140,000. The railroad's managers struggled to keep up with the volume by increasing the number of trains from 152 per day to 192 with little schedule accommodation made for the additional traffic.[6] Despite commercial telegraph facilities in many of its stations, the Eastern Railroad's managers did not use telegraphy for train dispatching. Instead, officials expected station agents and train crews to follow the railroad's strict operating rules and timetable.[7]

By the evening of Saturday, August 26, the high volume of travelers in both directions had thoroughly disrupted train service along the Eastern's line. Many trains were running off schedule as station agents in Boston struggled to dispatch them as soon as they were loaded with passengers.[8] Leaving the station, train crews had to contend with an intricate traffic pattern. Three miles north of Boston, the Saugus Branch, a single-track line, split off from the double-track main line at Everett Junction. Some northbound trains took the branch line, while others continued along the main line toward Lynn, Massachusetts.[9]

That Saturday evening, a series of trains proceeded north from Boston to Everett Junction. The first, a branch line train, waited at the junction for a delayed inbound train to enter the main line from the Saugus Branch. The delay forced the next three trains to stop behind the first one, since there was no siding at Everett.[10] Without permission to use the commercial telegraph facility at his station, the Everett station agent could not inform his superiors in Boston about the backup and delay on the main line. Near dark, the southbound train finally cleared the junction at Everett, and the first train exited the main line onto the branch line. The third train also exited the main line at Everett Junction. The second train, a local, continued along the main line. Behind it, the engineer of the Portland express, the fourth train in line, saw the train in front of him pull onto the branch line. Since the local main line train ahead of him was no longer in sight, the engineer of the express concluded that he had a clear track ahead of him.[11] As the local train continued along the main line, the express rapidly caught up to it. When the local stopped at Revere, the first station after the junction, the express crashed into it at full speed, its engineer having failed to see the red warning lights on the back of the local train because of the evening mist. In the collision and subsequent fire, twenty-nine people died and fifty-six received life-threatening injuries.[12]

As details of the collision became public, critics nationwide expressed shock and indignation that the Eastern's managers did not employ any form of telegraphic dispatching on the busy passenger rail line. The editor of the *Telegrapher*, a popular trade journal of the American telegraph industry, noted, "The efficiency of the telegraph has been so completely demonstrated by years of successful use on a great number of railroads that it is not easy to understand that any company with a considerable traffic would refuse to avail itself of it, or that it could be justified in such neglect." He hoped that Eastern officials would be held legally accountable for their irrational conservatism and be forced to implement telegraphic dispatching.[13]

A few weeks later, the Massachusetts Board of Railroad Commissioners, headed by Charles Francis Adams Jr., the great-grandson of President John Adams and a technocratic reformer, issued an official report about the Revere accident. Adams first gained national attention in the early 1870s for a book he had written with his brother Henry in 1871 about Jay Gould and Jim Fisk's financial plundering of the Erie Railroad in the 1860s. *Chapters of Erie* launched Adams's career as a railroad reformer and an antagonist to railroad speculators such as Gould. His knowledge of the industry and willingness to publicly call out bad actors engaged in irresponsible financial and operational management led to his appointment as head of the Massachusetts Board of Railroad Commissions. Now, he used the Revere accident to further his campaign for industry-wide reforms.[14]

The Massachusetts commissioners officially identified the lack of telegraphic communication between Everett Junction and the Eastern Railroad's Boston terminal as one of the key factors in the Revere accident. Adams noted, "Had [the telegraph], one of the oldest, most ordinary, and least expensive of appliances in operating railroads, been in use on the Saugus Branch upon the 26th of August, it is to the last degree improbable that a collision would have occurred." Adams's report noted that many other railroads in Massachusetts and throughout New England were just as poorly operated as the Eastern Railroad. He warned officials on these lines that they were exercising extremely poor judgment and inviting severe public scrutiny in the event of further passenger train accidents.[15]

Despite these public criticisms, New England managers continued to drag their feet and resist Adams's call to implement telegraphic dispatching. A year-end report issued by Adams and the Massachusetts commissioners criticized the lingering antitelegraph attitudes of the Massachusetts railroad establishment. The commissioners examined the causes of past railroad collisions in the state. They identified a number of general risk factors for collisions, including "imperfect [internal] regulations," "defective signals," and "want of telegraphic communications."[16] With respect to the Revere accident, they specifically cited "the want of a complete telegraph system which should keep the central office fully advised at all times of the exact position of each train on the railroad, and in communication with all of such trains at the several stations."[17] In other parts of the nation, railroad officials had already implemented dispatching practices based on these basic principles.

The commissioners also examined employee rule books to see how companies addressed common operating risks. They noted that under an ideal management model, "regulations for operating railroads should be as short and as few in number as possible, and so systemized that each employee at a glance can ascertain his own particular duties." Instead, railroad rule books were complex, confusing, and incomplete, particularly concerning telegraphy. The commissioners complained that "very few allusions to the telegraph, or to the duties of employees on receiving train dispatches were found in the whole collection."[18] The commissioners deemed the situation unacceptable and argued that poorly developed rule books perpetuated dangerous operating practices.

Adams and the other railroad commissioners sought to correct these dangerous operating practices by preparing a set of uniform rules for Massachusetts railroads. Since they lacked direct regulatory authority, board members hoped that companies would voluntarily adopt the new regulations. In preparing the rules, board members first submitted them to a committee of railroad managers. These industry representatives readily agreed with many of the board's suggestions, such as improving train brakes and installing "buffer platforms" between cars that would absorb the impact of collisions and keep the cars from being crushed, but the managers and the commissioners failed to see eye to eye on the subject of telegraphic dispatching.[19]

The commissioners' proposed rules contained three specific telegraph requirements that would modernize dispatching practices in the state. First, they insisted that all Massachusetts railroads operating single-track lines utilize telegraphy for controlling train movements. Railroad managers countered with a proposal that telegraphy be used "in aid" of time- and rule-based operating practices, as opposed to superseding these practices.[20]

Next, board members proposed that managers link stations by telegraph to establish a rudimentary block signaling arrangement by which trains would be separated by clearly defined intervals, or blocks, of space or time.[21] Railroad managers found this proposal "of questionable expediency," despite its widespread adoption in Britain, and argued that current operating rules already provided for space and time intervals without the need for expensive telegraph facilities and additional employees.[22]

Finally, the commissioners asserted that familiarity with telegraphy should be a condition for hiring station agents and other white-collar railroad staff. Industry representatives argued that only telegraph operators should be required to know Morse code. They feared that "disaster would be increased and aggravated by a dependence upon [the telegraph] in the hands of persons who would have only infrequent occasions for its use."[23]

These disagreements over telegraphy highlight a broader conflict between railroad managers and industry reformers in the postwar era. New England railroad officials, like American railroad officials in general, disliked outside parties trying to tell them how to manage their operations. They believed that they should be free to select equipment and managerial practices that complemented the operations of their individual railroads.[24] Railroad officials dismissed the commissioners' telegraphy recommendations, not necessarily because they believed that telegraphy had nothing to offer the railroad industry, but because they wanted to maintain firm control over managerial and technical changes on the lines. These changes often came with hidden costs and dangers and needed to be adopted incrementally, and with careful forethought.

The Massachusetts commissioners did not seem to understand why railroad officials opposed their recommended changes. They viewed managers' resistance to telegraphy as ignorance and stubbornness on the part of the officials. In their 1872 report, the commissioners vowed to make the telegraph "a necessary and recognized part of railroad machinery" through continued publicity about its value as a safety and efficiency tool. Board members concluded that managers who ignored their recommendations were doing so "at their own risk."[25]

The commissioners' well-publicized findings did not bring about the rapid changes that Adams and other reformers desired. New England railroad managers continued to publicly express doubts about the value of telegraphy for operational management. In his 1879 study *Notes of Railroad Accidents*, Adams recalled hearing a superintendent "gravely assure the [Massachusetts] railroad commissioners . . . that he considered [the telegraph] a most dangerous reliance which had occasioned many disasters, and that he had no doubt that it would be speedily abandoned as a practice in favor of the old timetable and running-rules system, from which no deviations would be allowed." Adams found these comments all the more grating, since they were uttered in the wake of the Revere accident, which he believed had demonstrated beyond a doubt "the impossibility of safely running any crowded railroad in a reliance upon the schedule." He concluded that only strict state regulation, or the threat of criminal prosecution, would force recalcitrant railroad managers in the Northeast to rapidly incorporate telegraphy into operational management.[26]

The Revere accident and subsequent state investigation highlight the continued reluctance of some New England railroad managers to employ telegraphy for operational management in the 1870s. This reluctance, however, must be viewed from a broader economic and technical context. The conditions that had compelled many railroad officials in the mid-Atlantic and the Midwest to adopt intensive telegraphic train management during the Civil War were not present in New England. Significant investments in double-tracking before the war enabled New England railroads to accommodate wartime passenger and freight traffic. New England officials did not have to manage rail lines in excess of two hundred miles long, as did officials on major trunk lines like the PRR, the Baltimore & Ohio, and the New York Central. Consequently, managers had little financial incentive to invest scarce operating capital in maintaining telegraph facilities or paying operators. They shrugged off Adams's demands for dramatic operational reforms and continued to make incremental changes to their detailed operating rules and strict scheduling practices. On the rare occasion when they had to use telegraphy for managing their rail operations, they did so on their own terms, not on the terms prescribed by railroad reformers.[27]

Block Signaling

The Revere collision highlighted the potential risks of operating busy commuter railroads without telegraphy, but as the 1864 Erie Railroad accident at Lackawaxen, New York, had demonstrated, the American system of telegraphic dispatching could prove just as dangerous. Adams argued that American railroads needed a better option: telegraphic block signaling. This new signaling method appealed to managers who were concerned about the inherent dangers of telegraphic dispatching, but its high construction and operating costs placed it beyond the reach of most railroads in the 1870s. Despite Adams's efforts to promote the new signaling concept, only a few well-capitalized companies could afford to experiment with block-signaling systems in the postwar era.

Adams used the Revere tragedy to promote industrial progressivism to railroad officials and the general public. His equipment-oriented approach to reforming railroad management largely structured public debates on railroad safety and regulation throughout the 1870s and 1880s.[28] He drew particular attention to telegraphic block signaling in the Massachusetts Board of Railroad Commissioners' 1872 report, and again in his 1879 book *Notes on Railroad Accidents*. He argued that the signaling method offered an alternative to poorly managed telegraphic dispatching practices and eliminated dangers posed by employee errors. At the same time, his efforts unintentionally highlighted the daunting technical requirements and the prohibitive operating costs of block-signaling systems.

When discussions of telegraphic block signaling first began to appear in American technical journals in the early 1870s, the practice was used by only a few American railroads. Under the practice, a train passed from one section, or block, of track to the next. By design, only one train would be allowed in a block of track at any given time. This feature prevented head-on and rear-end collisions and ensured efficient traffic flow when trains departed from their rigid schedules. Telegraphy enabled signal operators at the end of each block to contact each other when trains entered or exited their block. Using this practice, they could control train movements directly across the entire rail network.[29]

The Camden & Amboy Railroad (C&A) in New Jersey was the first to use block signaling. In March 1865, senior C&A officials approved the proposal of the company's vice president Ashbel Welch to establish "a system of telegraphic safety signals . . . by which a train passing one of the signal stations shall be informed whether the preceding train going in the same direction has passed the next signal and whether the track is clear."[30] Welch was intimately familiar with telegraphy, having served as the railroad's engineer at the time of the firm's first interaction with the Magnetic Telegraph Company in 1845, and had conducted extensive research on railroad signaling both in the United States and abroad during the subsequent decades. Following a physical breakdown in the late 1850s, Welch recuperated in England. Here he learned about the practice of block signaling on British railroads.[31]

British railroad managers had been experimenting with block signaling on heavily trafficked rail lines since the early 1840s. They used simple indicator telegraphs for controlling busy blocks of track near London, as well as major points of congestion, such as single-track lines through tunnels. British telegraphic block signal systems were labor intensive. Each signal tower had to be manned continually, and the required telegraph equipment was expensive. American railroads in the antebellum period could not afford the equipment and personnel costs necessary to operate block signals on rail lines that stretched hundreds of miles. As a result, they did not experiment with telegraphic block signaling before the Civil War.[32]

Welch's interest in block signaling was spurred by a major collision on the C&A's dangerously overcrowded main line between New Brunswick, New Jersey, and Philadelphia. Early on the morning of March 7, 1865, a northbound express passenger train rear-ended a local train north of Philadelphia. The local was running nearly two hours behind schedule when the collision occurred. Subsequent investigations showed that the local train's crew had failed to alert overtaking trains to its serious delay. The railroad's telegraphic dispatching practices also had failed to keep a safe time and distance interval between the two trains. Welch concluded that only block signaling would have prevented the collision.[33]

Welch and the C&A's superintendent, R. S. Van Rensselaer, quickly installed block signals along the railroad's busy main line between Philadelphia and New Brunswick. In an 1866 report, Welch described how the tower-mounted signals controlled train movements into sections of track. The signals consisted of a board with a black signal box in the center. A white panel appeared on the black signal box during the day, and a white light showed through the center at night indicating that a train could enter the block. Once the train began to pass the tower, the signalman lowered a translucent red panel, indicating that the block was occupied. Then he telegraphed the next station to report that a train had entered his block. The signal remained at stop until the signalman received notice from the next station that the train had exited the block. The new signaling protocol ensured that delays on the busy double-track line would not lead to rear-end collisions.[34]

By the early 1870s, the C&A, which was leased by the Pennsylvania Railroad in 1871, had extended its block-signal network between Philadelphia and Jersey

City.[35] Twelve signal stations regulated traffic between Philadelphia and New Brunswick. An additional thirteen stations controlled train movements between New Brunswick and Jersey City. In all, block signals provided for the safe and efficient movement of nearly sixty local and express trains a day in each direction on the line.[36]

Adams drew public attention to the C&A's block signals in the Massachusetts Board of Railroad Commissioners' 1872 report. He included a detailed study of block signaling by Franklin L. Pope, an expert in telegraphy and electrical science. Pope praised the C&A's telegraph-controlled signals because they provided for direct managerial control over train movements and ensured safe rail operations. Traditional telegraphic dispatching, on the other hand, provided only the illusion of safety, since collisions could still occur on telegraphically dispatched rail lines because of employee errors or unexpected train delays. Pope noted that safety and efficiency came with a price tag, however. Telegraph equipment for each station cost about $200 dollars per signal tower. The block signaling network required two dedicated telegraph wires, which cost roughly $175 per mile to construct. Each tower also required two signalmen for twenty-four-hour operation, costing about $40 per month. Altogether, the capital outlay for the block signals on the eighty-nine miles of track between Jersey City and Philadelphia would have been over $20,000 for telegraph equipment (notwithstanding the cost of the signal towers). Wages would have run nearly $2,000 per month.[37] Few companies in the early 1870s, other than the Pennsylvania Railroad, could afford such a significant and sustained financial outlay.[38]

Adams clearly favored block signaling over telegraphic dispatching practices, but as Pope's report demonstrated, the prohibitive operating costs of block signals placed them beyond the reach of most American railroads during the depression years of the mid-1870s. Railroad officials could not justify the cost of building block signals when many could not even pay interest on their bonds to investors.[39] Instead, they persisted in using the dangerous and poorly developed telegraphic operating practices that Adams abhorred. He continued to champion block signaling throughout the decade, but his efforts attracted far more attention from the general public than railroad officials. In the end, the Camden & Amboy and the Pennsylvania Railroad's block-signaling arrangement represented a highly innovative but largely impractical technical solution to the risks posed by the American system of telegraphic dispatching.

Railroads versus Western Union

Block signals were beyond the financial reach of most railroads in the 1870s, but many companies' contractual relationships with the Western Union Telegraph Company made incremental improvements to train dispatching practices nearly as difficult to accomplish. In the post–Civil War era, most companies did not have the operating capital available to construct and maintain their own internal telegraph networks. It was also unclear whether railroads could legally operate interstate telegraph subsidiaries.[40] Instead, most railroads received telegraph service from Western Union (WU) under contractual arrangements that provided WU with exclusive access to their rights-of-ways and saddled railroads with covering a significant portion of the maintenance costs for the poles and wires along their tracks.[41] The arrangement provided railroads with relatively inexpensive telegraph service, but WU prioritized profitable commercial message traffic over internal railroad correspondence and train orders, despite assurances to the contrary. Since WU's poor service often prevented managers from issuing train orders in a timely and efficient manner, they continued using the older American system of dispatching instead of developing more sophisticated train dispatching practices that depended on reliable access to telegraphy. Most railroad officials reluctantly accepted the consequences of their partnership with the commercial telegraph giant. Their concerns about operating costs trumped their interests in safer and more reliable rail operations. As a result, Western Union exercised an extraordinary degree of influence over operational management practices and indirectly inhibited technical and managerial reforms in the postwar era.

Soon after the Civil War, Western Union emerged as the dominant commercial telegraph company in the United States. In February of 1866, it acquired the United States Telegraph Company. A few months later, it merged with the country's remaining large telegraph firm, the American Telegraph Company. Following the mergers, Western Union exercised firm control over 37,380 miles of telegraph lines.[42] Though the company never established a monopoly over commercial telegraph communication in the United States, WU used its market dominance to keep other firms at bay. In particular, its exclusive right-of-way contracts with American railroads locked down over 95 percent of the nation's rail lines and posed a major barrier to entry for rivals.[43]

During the late antebellum period, railroad officials had seen inexpensive telegraph access for business communication as a great convenience, since few railroads could afford to license telegraph patents and build their own lines. Transmission delays rarely interfered with operational activities, since most railroads used the telegraph only occasionally to manage train movements. The financial marriage of convenience between railroads and Western Union gradually soured by the end of the Civil War, however. As many mid-Atlantic and midwestern railroads started employing telegraphic dispatching during the conflict, they began experiencing problems with their telegraph service. Managers found commercial lines clogged with business and government traffic and were unable to issue train orders in a timely manner. This seriously disrupted train movements on busy lines. Officials had to choose between minimizing telegraphic dispatching and investing prohibitively large sums of money to construct dedicated railroad telegraph lines. Given their financial constraints, most managers accepted the inevitable delays associated with Western Union service and tried to keep their trains moving as best they could.

Following the war, civilian managers continued to struggle with Western Union's poor service. In 1867, an anonymous writer in the *Telegrapher*, the official publication of the first commercial telegraphers' union, drew attention to railroads' "damaging disadvantage" compared with Western Union. The author noted that many railroad managers had been "considerably careless of their best interests" when they signed prewar deals with Western Union, allowing "undue advantage to be taken of their ignorance of the new element." He asserted that in exchange for granting many privileges to Western Union, "the [rail]road gets a wire, not always even good, never the best, and generally the worst—for which it pays a round price, besides giving the telegraph company right of way over its [route] for a longer or shorter period . . . the [rail]road in every instance paying the running expenses of its own wire." The writer concluded that WU officials had treated railroads unfairly in their quest to gain control over the American telegraph industry: "The telegraph managers took advantage of the necessities of the railroad and their ignorance of the whole business, and obtained from them these valuable privileges almost criminally."[44]

Railroads had undoubtedly received poor service from Western Union. They had indeed been given access to unreliable telegraph lines in exchange for valuable right-of-way contracts. More important, WU's stranglehold over railroad telegraph networks had inhibited important innovations in railway management. Instead of encouraging managers to develop new uses for telegraphy, WU officials tried to ration access to their facilities. Shortsighted senior WU managers consistently failed to recognize the financial benefits of having railroads as equal partners in their expansion across the nation.

The federal Telegraph Act of 1866 seemed to offer railroads an opportunity to challenge WU's exclusive control of their rights-of-way. The act granted telegraph firms the right to build lines along any post road, a provision that was interpreted to include railroad lines that carried the mail, if they agreed to let the federal government set rates for government telegraph business and also agreed that the federal government could buy the company after five years.[45] Encouraged by this act, a handful of railroads, including the Pennsylvania Railroad and later the Union Pacific and Central Pacific, opted to open their rights-of-way and telegraph poles to Western Union competitors that acceded to the provisions of the 1866 telegraph act. Railroad officials supported this action by arguing that ownership of the existing telegraph lines along their rights-of-way was unclear, since in many cases both the railroads and Western Union had contributed supplies and labor for their construction. This, in turn, prompted numerous lawsuits between WU and the rebellious railroads. Nevertheless, the majority of American railroads did not attempt to upend their inequitable relationship with WU.[46]

In the early 1870s, the *Telegrapher* again attacked Western Union's treatment of railroad companies and pointed out that poor telegraph service threatened the safety of all travelers. An 1871 article implied that WU bore part of the responsibility for the Revere collision because its policies had dissuaded New England railroad officials from experimenting with telegraphy.[47] A year later, an editorial praised the Pennsylvania Railroad for "owning, controlling and operating its own telegraph lines" and noted that this practice was the only way to assure safe railroad operations. Other companies had signed multi-decade contracts with Western Union that prevented them from making significant changes to their operating practices. Unfortunately, the editor of the *Telegrapher* noted, "the efficiency of [these] lines for railroad purposes is greatly impaired, and frequently the railroad business is subordinated to the more immediate requirement of the wires for private business." He further warned, "The time is fast approaching when railroads will be required . . . to establish and maintain an efficient telegraph system for the safety of their patrons, even more than for the facility afforded for working the [rail]roads systematically and effectively."[48]

The editor of the *Telegrapher* shared Adams's opinion that the telegraph was first and foremost a safety tool. While Adams strongly criticized railroad managers for failing to grasp this point, the *Telegrapher*'s editor argued that most managers could not treat the telegraph as a safety tool because they did not have sufficient access to it. WU's tight control over many companies' telegraph networks stood in the way of reform. The *Telegrapher* concluded that managers would never have sufficient access to railroad telegraph networks until they were willing and able to build and maintain their own lines.[49]

Western Union officials challenged this position. Instead, they tried to justify the status quo by arguing that railroads failed to use telegraphy effectively. "Observer," writing in the *Journal of the Telegraph*, Western Union's biweekly company magazine, noted that on most railroads, "the railroad wire is demanded more or less for the transmission of paid [railroad] business," rather than train orders. Likewise, he argued, if train dispatchers exercised "proper economy" in the transmission of orders, delays could be prevented. Managers should end the practice of having telegraph operators repeat orders back to dispatchers to ensure that they had been received correctly; "the time consumed by such useless repetitions ought to be saved in the interest of other business."[50]

These comments generated a variety of responses from dispatchers and commercial operators. "T. D." stated that he was "ready to adopt improvements" but failed to see how Observer's proposal was an improvement. In his view, the "only *safe*" method for dispatching involved double-checking train orders for errors by requiring telegraph operators to repeat the messages back to train dispatchers. Other correspondents poked fun at Observer for his ignorance of railroad operations. On the other hand, commercial telegraph operators applauded Observer. Correspondent "A" noted that he often had commercial messages delayed by railroad business, not the opposite. The epistolary debate continued for a number of months without a clear resolution.[51]

The controversy highlighted the fundamental problem caused by WU's control over many railroads' internal communication networks. Western Union officials failed to appreciate that telegraphy had become necessary for maintaining both safe and efficient operating conditions on many railroads. They viewed railroad telegraph service as a contractual arrangement that took limited resources away from their more profitable commercial operations. Railroad officials could not dramatically improve dispatching practices as long as Western Union impeded their ability to obtain prompt telegraph service.

Even railroads that did not depend on Western Union for telegraph services had run-ins with the powerful company. The Pennsylvania Railroad had built its own railroad telegraph lines in the mid-1850s and had obtained a Morse patent license, but it also opted to provide right-of-way access to Western Union after the firm acquired the Atlantic & Ohio Telegraph Company (A&O), whose lines ran along the PRR mainline. Since the PRR did not depend on WU for dispatching purposes, it held a distinct advantage over other companies. The strict provisions of the PRR's proposed 1865 universal telegraph contract, however, showed that the PRR's managers were not taking any chances in their business dealings with WU. Under the PRR contract, right-of-way privileges could not be transferred between telegraph companies without the permission of PRR officials. This clause prevented WU from gaining control over rival telegraph lines along PRR rights-of-way through acquisitions or mergers with smaller companies, as the telegraph firm had done with the A&O and many other firms with lines running along railroad routes during the 1850s and 1860s. Firms seeking access to the PRR right-of-way also agreed to furnish the railroad with unlimited access to their commercial lines for railroad business. Finally, the PRR limited all contracts to just ten years, giving the railroad additional flexibility in its business dealings with WU and other firms. Railroads that depended on Western Union services because they could not afford to license, construct, and operate their own telegraph facilities would never have been able to force such strict contract provisions on the commercial giant.[52] Two years later, the PRR permitted the Pacific & Atlantic Telegraph Company (P&A) to add a telegraph line along its right-of-way. The rival telegraph firm added cross arms to the railroad's telegraph poles, in some cases within feet of the Western Union line. WU officials sought to remove these cross arms from telegraph poles jointly owned by the railroad and WU. The legal case eventually ended up before a federal court, where the judges refused to grant an injunction against the P&A or recognize that its lines materially harmed the Western Union lines.[53] This decision provided an opening that financier and railroad investor Jay Gould would later employ in a sustained campaign against WU in the mid- to late 1870s.

Gould's efforts to create a viable Western Union competitor offered railroad officials hope that they could right the disadvantageous contractual relationships between their firms and WU. Gould had made major investments in western and midwestern railroads during the 1870s, but his primary interest was in taking control of Western Union. Gould's first assault on Western Union began in 1875 when he gained control of the relatively small Atlantic & Pacific Telegraph Company (A&P) and launched a rate war against WU.[54] At the same time, he waged a campaign to drive down WU stock prices by challenging the validity of a key telegraph patent and short selling WU stock. Gould also secured the rights to Thomas Edison's recently invented quadruplex telegraph, which would quadruple the capacity of existing telegraph lines, though it was not clear whether he would install the equipment on A&P lines.[55] Numerous railroad firms, including the PRR, signed contracts with the A&P for access to their rights-of-way.[56] The firm retained its independence for barely two years before Gould sold it to Western Union in August 1877 for a tidy profit. A&P continued to exist as a wholly owned subsidiary of WU, and railroad officials who had signed agreements with the firm hoping that it would help to disrupt the monopoly over commercial telegraph service must have been sorely disappointed by the outcome.[57]

Gould's second campaign against WU began in mid-1879, and once again he employed railroads as a key element in his fight for control of the telegraph giant. In this effort, Gould utilized an amendment to an Army appropriations bill passed by Congress in June of that year. The amendment was named for its sponsor, congressman and Gould lobbyist Benjamin F. Butler. Butler had commanded the Massachusetts volunteers who, while traveling to Washington, DC, at the start of the Civil War, were forced to commandeer the Annapolis & Elk Ridge Railroad line in Maryland after Southern sympathizers blocked their passage in Baltimore. Butler had emerged as a Western Union foe in the mid-1870s and advocated for competition within the telegraph industry based on the provisions of the Telegraph Act of 1866 that granted telegraph firms access to public highways if they agreed that the government could purchase the firm at fair market value.[58] Butler's 1879 amendment stated that railroad firms would have the right to operate commercial telegraph lines along their rights-of-way if they agreed to the terms of the telegraph act. The act nullified ultra vires legal concerns that railroads would be in violation of their charters by offering such a service. Gould used the Butler Amendment as the basis for establishing a new national telegraph firm, the American Union Telegraph Company, by repurposing railroad telegraph lines running along the rights-of-way of the railroads that he controlled and making partnership agreements with railroads that he did not control for access to their rights-of-way.

Gould found ready partners in a number of firms, including the Pennsylvania Railroad. PRR officials had been in the process of negotiating a contract renewal with WU when they received word of the Butler Amendment. The firm's board immediately resolved to halt negotiations, noting that the act "may so affect the powers and interests of [the PRR] as to require further consideration before the terms of the contract are closed."[59] The railroad eventually signed a ten-year, nonexclusive access agreement with American Union for an annual rent of $60,000.[60] The PRR offered the same terms to Western Union and its subsidiary, the A&P, but WU president Norvin Green refused, and the PRR's board voted in September 1880 to serve the A&P notice that it had twelve months to remove its telegraph lines from the PRR's right-of-way. The following month, the board issued the same notice to Western Union.[61]

Furthermore, Gould challenged Western Union's legal right to exercise exclusive right-of-way agreements with American railroads, which were arguably the basis for WU's industry dominance. The Supreme Court sided with Gould when he installed an American Union telegraph line along the Wabash Railroad's right-of-way. Gould controlled the Wabash and argued that WU's exclusive right-of-way contract with the firm was void, since the Wabash had agreed to the provisions of the 1866 telegraph act.[62] He employed a similar tactic on one of the western railroads that he had secured control of in 1873, the Union Pacific. Whereas the Wabash was a state-chartered firm, the Union Pacific and its sister line the Central Pacific Railroad possessed federal charters under the Pacific Railway Acts of 1862 and 1864. The 1864 act granted both railroads the authority to operate a telegraph line along their rights-of-way that would be used for commercial and railroad business. The act further stated that the firms could not restrict access to the line and had to transmit all messages tendered by customers. In 1869, both firms signed an agreement with the Atlantic & Pacific for a perpetual lease of the telegraph line in return for maintaining and operating it. Eight years later, the A&P came under WU control following Gould's first raid on the firm. After Gould created the American Union Telegraph Company in 1879, he demanded access to the telegraph line along the Union Pacific based on the terms of the 1864 Pacific Railway Act. The railroad's internal legal counsel (likely acting on Gould's orders) conveniently determined that the 1869 lease to the A&P was in violation of the 1864 act and moved to void the agreement and bring the telegraph line back under Union Pacific control. This precipitated a Western Union lawsuit seeking an injunction, which was granted. The federal government, in turn, declared itself to be a party to the suit, since it had chartered the railroad. The case slowly made its way through the courts and precipitated a congressional hearing in the mid-1880s regarding the obligations of federal land-grant railroads to provide right-of-way access to telegraph firms.[63]

Gould's telegraph gambit with American Union paid off. In addition to instigating a price war, he utilized his skillful market manipulation tactics to drive down WU's stock price and once again profited by short selling the stock. Western Union eventually caved and bought out American Union, securing Gould's control of the Western Union board in January 1881.[64] For independent railroads like the Pennsylvania, this sudden turn of events prompted an about-face. After giving WU twelve months to remove its lines from PRR property in October 1880, Pennsylvania officials were forced to resume negotiations with the firm in the spring of 1881 following American Union's purchase. By September 1881, they had a new twenty-year contract in place with Western Union affirming WU's right to access the railroad's right-of-way in exchange for a $75,000 annual rent. The only major benefit for the PRR was that the contract was nonexclusive, which gave them freedom to allow other telegraph firms onto their right-of-way for the same annual fee.[65]

For less wealthy or powerful American railroads that could not afford to operate their own telegraph facilities in compliance with the Butler Amendment and the 1866 telegraph act, Gould's scheme left them in the same position that they had been in before his assault on Western Union. They were dependent on an even more rapacious and powerful Western Union for all their telegraph service and had no leverage for imposing reforms on the commercial giant.

Ultimately, Western Union proved to be a poor business partner for American railroads in the post–Civil War era. The firm's tight control over many railroads' internal communication infrastructures hindered meaningful technical and operational reforms. In essence, market competition in the telegraph industry obstructed efforts to improve railroad safety. Instead of expanding telegraph services to keep pace with growing traffic demands, WU's restrictions forced managers on poorly capitalized railroads to limit access to the communication tool. This furthered their dependence on the older and less reliable American system of dispatching.

The American System Prevails

Railroad managers accepted the shortcomings of the American system because they had no viable alternatives. It enabled them to stretch their limited financial and telegraphic resources to deal with the increased operational tempo of the late 1860s and early 1870s. Even though the poorly developed methodology sacrificed safety for efficiency by placing tremendous responsibility in the hands of train dispatchers and telegraph operations, most managers felt that the ends justified the means. While officials gradually came to recognize that the American system was deeply flawed and permitted a high rate of operational accidents, few had the organizational, technological, or financial resources to implement meaningful reforms such as building their own telegraph lines or developing more sophisticated telegraph-based operating practices. Out of necessary, they doggedly clung to the American system and loudly defended it against outside critics.[66]

American railroad officials' attachment to the American system with its awkward combination of time-, rule-, and telegraph-based operation practices is illustrated in a debate that appeared in the pages of the *Railroad Gazette* in the early 1870s. In 1872, "Hindoo," a British colonial official visiting the United States, wrote a series of articles on railroad management for the *Railroad Gazette*, one of the key trade journals for the American railroad industry and an open forum for discussing railroad equipment and management. Its editor routinely printed letters on telegraphy and other managerial issues from train dispatchers and railroad officials to highlight problems within the industry and advocate for reform.[67] In a February 3 article, Hindoo dissected the American system and compared it to managerial practices in British colonial India. He noted that Indian railroads almost never suffered from "butting," or rear-end, collisions, although many were single-tracked just like American lines. Hindoo ascribed this difference to Indian block signaling practices adopted from the British railroad industry by colonial railroad officials. Under the Indian method, each station served as a "blocking" point. Before a train could leave a station, the telegraph operator had to receive word from the next station that the line between was clear. Then the operator issued a clearance card to the train crew and allowed them to proceed. The dispatching protocol's most beneficial feature, Hindoo pointed out, was that it eliminated ambiguity and the possibility that dispatchers or other employees could make errors while transmitting train orders by telegraph and cause collisions.[68]

Hindoo contrasted Anglo-Indian train management practices with American telegraph-based dispatching, which he asserted were "theoretically good" but "in practice . . . unsafe [and] unreliable" because they put too much authority in the hands of one man, thus sacrificing safety for economy. Hindoo asserted that "few superintendents know the number of mistakes that are made, the number of hairbreadth escapes, the delays to trains through inattention or want of judgment [on the part of dispatchers]." He concluded that American managers would benefit by looking to Indian operational practices for guidance.[69]

Railroad Gazette readers responded quickly. Two weeks later, a correspondent writing under the nom de plume "X" challenged Hindoo's conclusions. X framed his argument by noting: "The system described [by Hindoo] seems to be the usual British system, which is doubtless safe. No man will deny that. But the question comes up: Can all be as safe by any other system which may be more economical?" He argued that the American system of telegraphic train dispatching was far more efficient and cost-effective than Indian and British block signaling practices. In addition, it was just as safe if "carried out faithfully" by employees.[70]

That issue of the *Gazette* also contained a response by "America," who charged that Hindoo had drawn his conclusions "from a defective execution of the [American] system and not from any fault of the system itself." Like X, America argued that American telegraphic dispatching practices provided safe operating conditions at a fraction of the cost of those employed in India. He concluded, "Accidents that result from the movements by telegraph are not so much the fault of the system as they are violations of orders and rules" by employees.[71] Thus, in his view, railroad personnel, regardless of the dispatching protocols they observed, bore ultimate responsibility for keeping trains moving efficiently and safely.

Hindoo responded to his critics by asserting that the dozens of relatively minor head-on and rear-end collisions on American railroads each month proved his point that telegraphic dispatching in "theory, as well as [in] practice, is far from perfect."[72] Like Charles Francis Adams Jr., he argued that the American emphasis on economy was misplaced and dangerous. "Cheap locomotion is a good thing, but safe locomotion is better." Hindoo concluded that if managers failed to appreciate the consequences of their penny-pinching ways, they would face increasing calls for regulation from people outside the railroad industry who had little understanding or appreciation of railroad management. If that happened, Hindoo warned, the "science of management" would never improve.[73]

The unusually lengthy written debate between Hindoo and his American critics continued largely along these lines in the *Railroad Gazette* through the spring of 1872. Correspondents continued to challenge Hindoo's assertion that British-derived block-signaling practices would work on American railroads in the place of telegraphic train dispatching. They further contended that the American system offered efficiency at a reasonable cost. One correspondent joked that under Hindoo's methodology, trains would "get flat-wheeled waiting at stations for [train] orders."[74] Hindoo continued to assert that the American system was really a system in name only and provided no checks against employee mistakes: "No system can be safe if it leaves important movement altogether in the hands of one man. The very best man is liable to err." He pressed X and America to explain why collisions frequently occurred on American railroads if the American system of telegraphic train dispatching had reached such a state of perfection.[75]

Throughout the debate, Hindoo's critics maintained that the American system worked, despite numerous accidents. X pointed out that many of the collisions Hindoo referenced in his writings had taken place on small lines that did not even use telegraphic dispatching. He argued that the efficiency of domestic railroads illustrated the viability of American dispatching practices: "Where you see 1,800 or 2,000 trains per month, handled on a single track under any system, without delay or accident, there is some evidence of safety and efficiency." At the same time, X acknowledged that financial and operational limitations restricted managers' abilities to improve their operating methods: "That there is room for improvement in our machinery we will admit; but that we must make our organization cumbersome and disregard expense we do not believe. Safety is only a relative term at best."[76]

A few *Railroad Gazette* contributors acknowledged Hindoo's concerns that the American system placed too much responsibility in the hands of the individuals that made up the dispatching chain. One correspondent recalled a twelve-year-old telegraph operator being sent to serve as the night operator "at a station of considerable importance."[77] Another stated more generally that "some of our railroad managers have fallen into the error of supposing that it need take but a few months to educate a boy, whose only qualification is the skill to send a telegraph message correctly, to perform the most responsible of all duties upon a railroad, viz., that of dispatching trains."[78] These members of the railroad community recognized the futility of cost-cutting measures that placed inexperienced or unqualified personnel in positions of power, where they could do great financial damage to their employers. Yet, the *Railroad Gazette* correspondents could not offer concrete suggestions about how to remedy the inherent flaws in the American system.

Without reaching any conclusions, the debate sputtered to an end in June. Hindoo would not accept management practices that did not "recognize the frailty of human minds, and which [did] not provide a check against such casualties."[79] Many of his critics, on the other hand, found his ideas impracticable for financial and organizational reasons. This fundamental disagreement highlights the difficulties faced by reformers in the early 1870s. Despite public calls for technical and managerial change, such as adopting telegraphic block signaling, few railroad officials had the resources available to implement meaningful reforms. They depended on the American system because it was inexpensive and it kept the trains running through its hybrid time-, rule-, and telegraph-based operating practices. While block signaling was safer, it was much more costly and required significant changes in how managers operated their lines. Thus, its appeal was extremely limited.

In the midst of the debate, a few industry reformers stepped forward to suggest new ways to address the problems caused by the American system. On March 16, the *Gazette* reprinted an article from the *Telegrapher* in which "Mac," a train dispatcher, complained that only the most rudimentary dispatching practices were universally observed on American railroads. Specific dispatching rules differed greatly, increasing the risks of dispatching errors when dispatchers and telegraphers changed companies. Mac called for a national convention of chief train dispatchers to address these inconsistencies and develop a uniform set of telegraphic dispatching practices for American railroads.[80] W. W. Wells, superintendent of telegraph for the St. Louis, Iron Mountain & Southern Railway, seconded the call for a convention of dispatchers to establish "the best and most uniform system of train dispatching."[81] Despite their enthusiasm, there is no evidence that a train dispatcher convention ever took place. While these correspondents tried to promote reform from the bottom up, their proposals failed to generate action on the part of dispatchers and telegraphers.

Other industry reformers tried to institute change from the top down. In May 1873, the Committee on the Best Method of Dispatching Trains by Telegraph for the short-lived Railway Association of America issued a report that illustrated the shortcomings of American operating practices. The committee complained that "the important subject of train dispatching has not yet received that careful study and dissection which are necessary to the attainment of a high state of utility." They concluded that few managing officials were willing to address the serious flaws in the American system. Managers blamed employee mistakes for most accidents and called for greater discipline, instead of investing more financial and organization resources into reforming telegraphic dispatching practices to eliminate ambiguity and the likelihood of communication errors.[82]

The committee acknowledged that the American system offered "economy, simplicity, and the facility of moving trains along the [rail]road," and they did not dispute that it was "immeasurably superior to the old style of working by schedule only." Members argued, however, that the large number of collisions on American railroads compared to British rail lines raised doubts about "whether safety and economy are characteristics of the system."[83] To solve these operating problems, committee members offered some practical suggestions. They recommended improving the uniformity of train rules and timetables to reduce companies' dependence on telegraphic dispatching. Members also spoke positively about block signaling in their report but acknowledged its prohibitive costs for many companies. Ultimately, the industry committee, like Hindoo and Adams, concluded that railroad officials desperately needed to improve their train operating practices through improvements to their organizations and equipment. But the committee's arguments could not outweigh the dismal economic realities that many railroad officials faced in the mid-1870s.[84]

Conclusion

Throughout the 1870s, American railroad officials struggled with the legacy of their ad hoc, late antebellum, and wartime managerial reforms. While telegraphy had enabled railroads to manage the traffic boom of the previous decade, the poorly implemented mix of newer and older operating practices significantly increased operating risks. In the postwar era, most officials struggled to find a middle ground between the prohibitively expensive technical reforms promoted by Charles Francis Adams Jr. and the antitelegraph rhetoric of some New England railroad officials. Their efforts were impeded by Western Union's tight control over their telegraph infrastructure, which limited their ability to improve dispatching practices though incremental changes. Following the Panic of 1873, few railroads had the necessary financial resources to build their own telegraph networks or implement major operational reforms. Similarly, Western Union, under assault from Jay Gould during this decade, did not wish to expend its resources on expanding unprofitable railroad telegraph services. Instead, managers accepted the limitations of the American system and tried to minimize operating risks through less costly internal campaigns to improve employee discipline and responsibility. These measures became less effective as railroads continued to expand their service and consolidate into larger networks as the nation slowly pulled out of its economic depression at the end of the 1870s.

The Struggle for Standards

> I consider it not nearly as important what kind of signal is used as to have uniform signals on all [rail]roads.
>
> O. S. Lyford, General Superintendent, Chicago & Eastern Illinois Railroad, 1883

The decades following the Civil War were a time of unprecedented, and often ill-considered, growth for American railroads. Despite a few short and painful economic contractions in the early 1870s and 1880s that slowed expansion and forced a number of firms into receivership, railroads across the nation increased dramatically in size and traffic density during this era. Large, well-capitalized firms such as the Pennsylvania Railroad and New York Central acquired control over many smaller rail lines and attempted to integrate them into broader operating networks. Railroads also established strategic alliances with other firms and gained access to their tracks and facilities. In both cases, senior managers struggled to resolve problems created by different railroads' incompatible and sometimes contradictory operating rules and telegraphic train dispatching practices. These incompatible operating standards led to numerous train accidents and near misses.[1] Beginning in the mid-1870s, officials with the PRR devoted great attention to standardizing operating and telegraphic train dispatching practices across the firms' many divisions and updating the company's outdated employee rule book. While officials did not radically overhaul the PRR's operational management practices, they did develop a familiarity with railroad operating protocols and train dispatching practices that allowed them to exert significant influence over broader efforts to establish industry-wide standards for American railroad operations in the mid- to late 1880s.

In 1883, the General Time Convention (GTC), a railroad industry association consisting of senior officials from lines throughout the nation, began to consider the growing problems posed by inconsistent operating and dispatching standards. Working through the GTC's Uniform Train Rules Committee, PRR managers successfully promoted their firms' own train dispatching and operational management practices as the basis for the committee's new universal operating code, the Standard Code of Train Rules. PRR officials' actions were not warmly received by some GTC participants, but PRR and allied railroad officials dominated the acrimonious debate over the implementation of the rules and succeeded in gaining official sanction for the new standards in 1890.

American train dispatching and railroad operating standards were shaped by specific organizational, political, and technical influences from both within and outside the railroad industry. The standard-setting process was driven by the coercive actions of a powerful minority within the railroad community, rather than through consensus-building efforts by the industry as a whole. Consequently, the Standard Code of the GTC reflected PRR officials' particular vision for the industry. Many member firms eventually adopted the new operating standards, but they were never universally accepted, as had been the hope of GTC leaders at the beginning of the process.[2]

Interfirm Cooperation in the Civil War Era

American railroad development throughout much of the 1830s and 1840s reflected an extremely localized focus. Political and economic provincialism hindered the development of regional rail networks. While railroad managers openly shared technical and operational information with each other through interpersonal exchanges and industry publications, officials saw little need to coordinate equipment standards and operational practices between their geographically isolated firms.[3]

Only in the late 1840s, as traffic density increased significantly across the region, did managers in New England first begin to consider the advantages of coordinating technical and operational practices and developing uniform regional standards.[4] During the 1850s and 1860s, the American railroad network expanded dramatically in length and traffic volume. Numerous firms completed trans-Appalachian trunk lines connecting the Atlantic seaboard with the Ohio River Valley. A growing web of short "feeder" lines branched off from these trunk lines and dramatically increased traffic flow across the networks of major firms such as the Baltimore & Ohio, New York Central, and Pennsylvania Railroad. As competition intensified between major railroads for freight and passenger traffic, particularly along east-west trunk routes, managers looked to interfirm associations to promote corporate "harmony." Officials organized numerous, short-lived railroad conventions in the early 1850s in efforts to resolve rate wars and address operational and technical uniformity concerns. While these voluntary, and ultimately temporary, organizations attempted to generate group consensus to promote new industry standards, railroad officials' efforts were not successful. "In a half dozen cases, the convention delegates adopted a formal name for their association and intended it to be permanent, but in every case the association seemed to have disintegrated shortly thereafter."[5]

Following the Civil War, railroad officials from across the country once again attempted to organize an industry-wide association to address pressing political, economic, and technical concerns. During the war, the federal government had asserted its regulatory power over northern railroads. A congressional act in 1862 authorized the president on behalf of the War Department to seize control of civilian railroads and impose military authority over railroad personnel should wartime circumstances necessitate these actions.[6] In addition, the federal government imposed numerous taxes on railroad operations to help pay for the costly war effort. During the summer and fall of 1866, railroad officials assembled as the National Railway Convention (NRC) to discuss these new political and financial circumstances and establish a permanent railroad bureau to lobby the federal government on behalf of the industry and promote technical and operational standards to the railroad community.[7]

In addition to discussing issues relating to competition and profitability, NRC members expressed an interest in addressing technical matters relating to railway signaling and train dispatching. Officials created the Committee on Safety, Signals, and Regulations to evaluate operating practices on member railroads and suggest "best practices" to avert collisions and other types of operating accidents. In an October 1866 report to the NRC, committee chairman Ashbel Welch of the Camden & Amboy (C&A) Railroad of New Jersey offered a scathing critique of the lax operating standards observed by railroad officials across the United States. He blamed improper and inconsistent signaling practices for causing the majority of train accidents. Welch asserted that signals should show locomotive engineers when they could proceed safely down a line, not warn them of unknown dangers that might lie ahead. While the semantic difference might seem minor, in reality, Welch's proposal would represent a major change in how railroad operating practices were conceived and implemented. On busy double-track lines where trains generally ran in a single direction along each track, Welch cited inadequate signaling as the most frequent cause of rear-end collisions. Instead, he promoted the C&A's innovative positive train control practices as the solution to this problem. The C&A's block signals allowed signalmen to communicate train locations to each other via the telegraph and control train movements through each block of track. Since only one train would be allowed in each block at a time, it would be physically impossible for trains to collide. As an added benefit, block signaling also promoted more efficient train movements across the C&A's busy network, since it reduced waiting times at stop signals. Welch concluded his report to the NRC by pleading for uniformity in all railroad-operating practices: "Signals and safety regulations should be uniform for each [rail]road, and as far as possible for all [rail]roads, especially for those connected with each other. The apparatus and mode of working should be minutely described and carried out with precision. As little as possible should be left to discretion. Emergencies should be provided for by rules deliberately made."[8]

Welch's recommendations highlighted the growing operational risks that railroads faced in the postwar era. During the war, firms had begun to run "through" trains across each other's lines. Inconsistent signaling and train dispatching practices created confusion among train crews operating on shared lines and increased the risk of collisions and other types of accidents. At the end of the war, major trunk lines began to acquire smaller feeder lines to expand their rail networks. These formerly independent lines had observed a multiplicity of operating rules and regulations, which further muddled operating standards across the growing networks. In his 1866 NRC report, Welch warned officials that these discrepancies must be resolved before they disrupted rail operations and undermined public confidence in the rail industry. He emphasized that new standards must be clear, specific, and hierarchical. The fewer judgment calls made by train crews and station agents the better. Welch's implied warning was clear in the report: if railroad officials could not reach a consensus about uniform operating practices, they risked having new safety and operating standards imposed on them by state and federal lawmakers.[9]

Despite efforts by NRC delegates, the convention did not follow through on any of Welch's prescient safety recommendations. Members also failed to fund a railway bureau, which might have provided a vehicle for interfirm cooperation in developing uniform operating practices and safety standards. Instead, managers were left with no overarching guidance as they struggled to implement operating standards on their individual railroad lines. The NRC faded away after 1866, having failed to accomplish its goal of representing the technical and managerial concerns of the railroad industry. Over a decade passed before railroad officials once again convened to discuss uniform operating standards on American railroads.[10]

Internal Standardization

The absence of uniform, interfirm operating standards hindered the development of safe and efficient rail networks in the post–Civil War era, but individual railroad firms bore much of the blame for the situation. Rail networks cobbled together from many smaller firms strained under the burden of inconsistent internal operating practices, which led to high accident rates as train crews encountered unfamiliar rules and regulations when they crossed divisional boundaries between formerly independent rail lines. Responding to this growing problem, Arthur Wellington, editor of the *Railroad Gazette*, echoed Ashbel Welch's earlier warnings: "The present condition of [the operating] department of railroad management is that of complete confusion. Many [operating] codes lack the element of consistency in themselves, and most of them may be characterized as mere aggregations of orders issued at various times, by all sorts of men, and under widely varying circumstances. . . . Some of the most important [rail]roads in

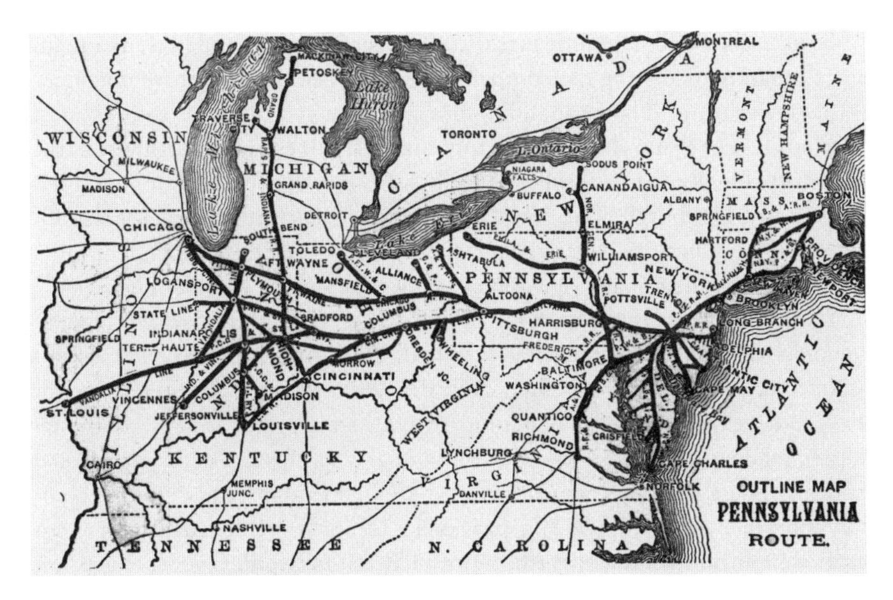

The Pennsylvania Railroad expanded dramatically during the 1870s and 1880s as the firm acquired numerous formerly independent railroads in the East and Midwest, as well as built out its core line between Philadelphia and Pittsburgh. As this 1892 map illustrates, the railroad's network eventually stretched from New York to St. Louis and Chicago. Formulating and imposing consistent telegraphic dispatching and operating standards across the network proved to be a key challenge for the railroad's management during this era. This led to the creation of the Association of the Transportation Officers as an internal standards-setting body. *The Schwantes-Greever-Nolan Travel and Transportation Ephemera Collection, St. Louis, MO*

the country are conducted with very little system."[11] Senior managers had to resolve network-wide inconsistencies and implement new operating codes to alleviate this chaotic and dangerous situation. For large trunk lines, this meant promoting internal standardization to manage operational activities taking place along thousands of miles of track.

The Pennsylvania Railroad led the way in developing consistent internal standards. PRR officials had engaged in a massive system-building effort prior to the economic downturn of the mid-1870s. By 1874, they had acquired dozens of formerly independent railroads and gained control over a network stretching from New York City to St. Louis.[12] Because of this unprecedented growth, PRR managers faced the challenge of developing consistent train operating practices across its entire rail network sooner than other rail officials. The practical experience that PRR officials gained in the late 1870s and early 1880s later enabled them to assume dominant leadership roles as industry-wide reformers.[13]

In 1874, the PRR's longtime president J. Edgar Thomson and his senior managers began reorganizing the company's new acquisitions into three major administrative units. The Panhandle lines encompassed divisions southwest of Pittsburgh. The Pennsylvania Company controlled lines to the northwest of the Steel City. Finally, the Pennsylvania Railroad Company operated the core PRR lines east of Pittsburgh. The new administrative structure enabled senior officials to impose firm managerial control over these formerly independent lines, as well as implement major operational changes across the entire rail network.[14]

During the reorganization, PRR officials chose not to interfere with operating practices on the lines they had acquired. Only after the firms had been integrated organizationally and financially did senior managers begin looking for ways to develop and implement company-wide standards.[15] In 1875, divisional superintendents initiated a series of semiannual meetings to discuss operational reforms. These meetings, which included PRR general manager Frank Thomson and other corporate officials, provided a means for senior managers to coordinate reform efforts across the entire PRR rail network. Over the next few years, participants in the superintendents' meetings organized standing committees to investigate signaling practices, train rules, and other important operational issues.[16]

In 1879, participants in the semiannual meetings founded the Association of the Transportation Officers (ATO).[17] The new association served as a permanent advisory body for the company's board of managers. It was tasked with "the general improvement of the Pennsylvania Railroad Company's Service."[18] The ATO included divisional superintendents, assistant superintendents, and other managerial and engineering staff officers from the PRR and its many affiliates. It also incorporated the standing research committees that had been previously organized by company superintendents. General Manager Thomson chaired ATO meetings, which signaled to observers that the semi-independent body would play a direct role in shaping company policies.

Before each meeting, ATO members submitted operational questions, such as the best color for railroad signals, to the secretary of the ATO. He referred these questions to the various research committees for analysis. They later reported their findings at quarterly meetings and accepted comments and criticisms from the general ATO membership. In all, the new organization offered a practical forum for developing and implementing company-wide operating practices.[19]

One of the first issues that the ATO addressed was updating the PRR's operating rules. At a January 1880 meeting, Superintendent John A. Anderson suggested that members convene a committee to examine the shortcomings in the 1874 rule book. At the time of its approval, the 1874 rule book had significantly modernized the PRR's operating practices by incorporating detailed rules for dispatching trains telegraphically. PRR managers had acknowledged the necessity of telegraphy for daily operations and sought to develop standardized practices for operating trains safely and efficiently under its control. Despite these improvements, Anderson believed that the dispatching techniques mandated by the 1874 rule book had become obsolete.[20]

Anderson had been keenly interested in train dispatching since the early 1870s. As superintendent of the Belvidere Division of the Pennsylvania Railroad, he had experimented with different telegraphic dispatching protocols in hopes of finding a safer and more efficient means of issuing orders to train crews. In 1873, the *Railroad Gazette* published his thoughts about dispatching in an essay titled "Rules for the Movement of Trains by Telegraphic Train Orders." Anderson continued experimenting with different dispatching methods. By 1880, he was arguably one of the most knowledgeable ATO members on the subject of train orders and dispatching practices.[21]

Given Anderson's expertise, it is not surprising that he helped guide the ATO committee tasked with revising the 1874 rule book. He was particularly interested in seeing duplicate, or double-order, dispatching instituted across the entire Pennsylvania rail network. Double-order dispatching required that "all trains concerned in the execution of a specific movement should receive the order in the same words."[22] In practice, each train crew would receive a memo that listed both their new schedule and the new schedules of any other trains affected by the order. Under the older single-order dispatching model, train crews had received schedule changes for only their train. As a result, they did not know how other trains on the line would operate under the revised schedule. Duplicate orders improved safety and efficiency by providing an additional safeguard against the possibility that train dispatchers might mistakenly issue conflicting orders to train crews, leading to delays or collisions.

Despite Anderson's influence, the new 1882 PRR rule book contained few major revisions. Telegraphic train orders, which had been termed "Special Orders" in the 1874 book, now were referred to simply as "Train Orders."[23] The change reflected the increasing normalcy of telegraphic dispatching on the PRR. More importantly, however, the new rule book failed to mandate double orders explicitly. Dispatchers on each division were allowed to employ either single or double orders according to their individual preference.[24]

Even though Anderson failed to radically overhaul the PRR's rule book, his pioneering work as an ATO member helped bring his train dispatching reforms to the attention of the broader railroad community. In October of 1882, the *Railway Age*, a national trade journal similar in scope and content to the *Railroad Gazette*, published the first in a series of articles by Anderson on train dispatching. In the introduction to his first article, Anderson argued, perhaps in reference to his service with the ATO, that "the value . . . of the telegraph as a railway appliance is not as yet sufficiently realized, and hence its capabilities for usefulness have not been developed to an extent commensurate with its importance."[25] In later articles, Anderson promoted double-order dispatching as the safest and most efficient means of directing train movements in the United States.[26] Anderson's essays became the basis for a book published the following year, titled *The Train Wire: A Discussion of the Science of Train Dispatching*.[27]

Anderson's knowledge of train dispatching, nurtured by his participation in the PRR's superintendents' meetings, and later the ATO, secured him a role as an industry expert on telegraphy and railroad management. Along with Anderson, other ATO participants also emerged as leaders in a renewed drive to develop uniform, interfirm operating standards for American railroads in the mid-1880s. As managers from across the country once again began to consider national standards for railroad operating practices, PRR officials used their recent internal standards-setting experiences to direct the broader national standards-setting effort.

Universal Standardization Revisited

After the National Railway Convention failed to achieve interfirm cooperation and managerial consensus in the 1860s, specialized rail industry trade associations emerged to resolve pressing technical and managerial issues. The Master Car-Builders' Association, composed of superintendents from most major US railroads, played a direct role in developing standard freight car designs and procedures for exchanging traffic between different lines in the 1870s and early 1880s. Other trade groups promoted integration by addressing specific economic and technical reforms. The most influential trade association of the era, the GTC, accepted the challenge of developing and implementing industry-wide operating and dispatching protocols.[28]

Under the influence of current and former Pennsylvania Railroad officials, the GTC produced the first uniform set of operating rules for American railroads in the late 1880s. Known as the Standard Code, these rules specifically addressed many of the dangerous and inconsistent operating practices that industry reformers had complained about throughout the 1870s and 1880s. Not all GTC members cared for the Standard Code. Some accused PRR officials of seeking to impose Pennsylvania Railroad operating protocols on the entire rail industry without concern for individual railroads' operating costs and organizational requirements. While the GTC could not force members to adopt the new code, some officials opposed to the code feared that the PRR would use its size and financial strength to coerce smaller firms into accepting a "one size fits all" approach to operational management. Thus, rather than the result of industry consensus, the development of the Standard Code revealed fault lines in the American rail industry and highlighted the anxieties and concerns that some officials felt toward powerful, dominating companies like the Pennsylvania Railroad.

The need for uniformity in railroad operations became increasingly clear in the 1870s and 1880s. Despite vocal campaigns by public safety advocates such as Charles Francis Adams Jr. to promote automatic signaling and train control devices, most American railroads continued to move freight and passenger traffic across their rail networks using manual telegraphic dispatching practices.[29] On each rail line, a small army of telegraph operators reported train locations to divisional dispatchers as trains passed by their stations. Dispatchers monitored train movements across entire operating divisions and periodically telegraphed train orders to station agents to pass along to train crews when delays threatened to disrupt the daily schedule. On single-track lines, train orders typically modified timetable schedules and passing points for oncoming traffic to account for unexpected delays or backups. On double-track lines, dispatchers typically used train orders to maintain safe operating distances between traffic moving in the same direction, and sometimes directed faster trains to temporarily switch tracks to bypass slower traffic. The train dispatching process was fraught with risks. Simple mistakes by dispatchers or telegraph operators could set two trains on a collision course with no recourse other than alerting wrecking crews to the pending accident. Despite these risks, officials did not feel compelled to replace manual dispatching practices with expensive and unproven automatic signaling and train control devices.[30]

While railroad officials resisted efforts by public safety advocates to impose new train control tools on their firms, many recognized that operating practices in the United States were extremely inconsistent between different firms. Inconsistencies might not pose serious problems for experienced train crews operating on familiar routes, but they could prove to be a significant source of danger as firms began sharing access to sections of track and train crews began running trains across unfamiliar rail lines belonging to other companies. For many industry reformers, the increasing rate of train collisions in the late 1870s and early 1880s highlighted the growing danger that inconsistent operating standards posed to the American rail network. By the early 1880s, reform-minded officials recognized that interfirm cooperation would be essential to address these inconsistencies and develop uniform, nationwide railroad operating standards.[31]

The GTC provided an ideal vehicle through which reform-minded railroad officials could promote interfirm cooperation and operational uniformity. Senior railroad officials in the Northeast and Midwest had originally organized the GTC in the early 1870s to help member railroads coordinate biannual schedule changes. A sister organization, the Southern Railway Time Convention (SRTC), served the same function for southern rail lines, which, until 1886, operated under a five-foot gauge instead of the four-foot, eight-and-a-half-inch gauge that railroads elsewhere in the United States used. Interfirm schedule coordination had become increasingly necessary by the 1870s, as companies began running long-distance trains across lines belonging to partner railroads. Twice-yearly GTC and SRTC meetings, often held in tandem, provided forums for senior officials from different companies to address schedule changes along with other matters relevant to daily operations.[32]

The General Time Convention first gained industry prominence as a result of its efforts to implement national standard time zones in the early 1880s. The New England Association of Railway Superintendents' efforts in the 1840s to create a standard railroad time zone had met with limited success. In the decades that followed, most American railroads continued to base their internal time on the civic time observed in the major cities in which they were headquartered. As Alexis McCrossen notes: "Ten U.S. cities set the time for more than half the hundreds of railroads running in 1874. In 1883, just eight American cities set the time for two-thirds of the nation's railroads; the remaining third followed forty-one different civic time standards."[33] For rail stations not located in major cities, multihanded clocks displaying the local civic time and the railroad time observed by the firm operating the rail line were a common sight. Beginning in late 1881, GTC officials sought to remedy this timekeeping nightmare and the inherent problems it caused for rail firms as they increasingly tried to coordinate their operations and share their tracks. The GTC was in a much more powerful position to implement new standard timekeeping practices than early regional railroad associations had been. Its membership included railroads from across the nation that could institute standard time zones along thousands of miles of rail lines without consulting civil authorities, creating de facto national time zones.

William F. Allen, who edited the *Travelers Official Railway Guide* and served as secretary of the GTC, took the initiative in promoting the benefits of standard time zones to member railroads.[34] Time zones would enable railroads to coordinate regional train schedules more effectively by eliminating the confusing multiplicity of local time zones observed by different companies, thus reducing "the present uncertainty to comparative if not absolute certainty."[35] Allen initially sought to promote standard time at the GTC's spring 1882 meeting, but a rate war forced the indefinite postponement of the meeting. In the months leading up to the organization's fall 1882 meeting, Allen published numerous reports on standard timekeeping, and Congress passed a law authorizing the president to call for an international convention to establish a prime meridian for timekeeping purposes. At the fall 1882 GTC meeting, members briefly set aside their differences regarding rates and took note of Allen's recommendations to preempt any further congressional action regarding timekeeping by establishing uniform time standards favorable to the railroad industry. They

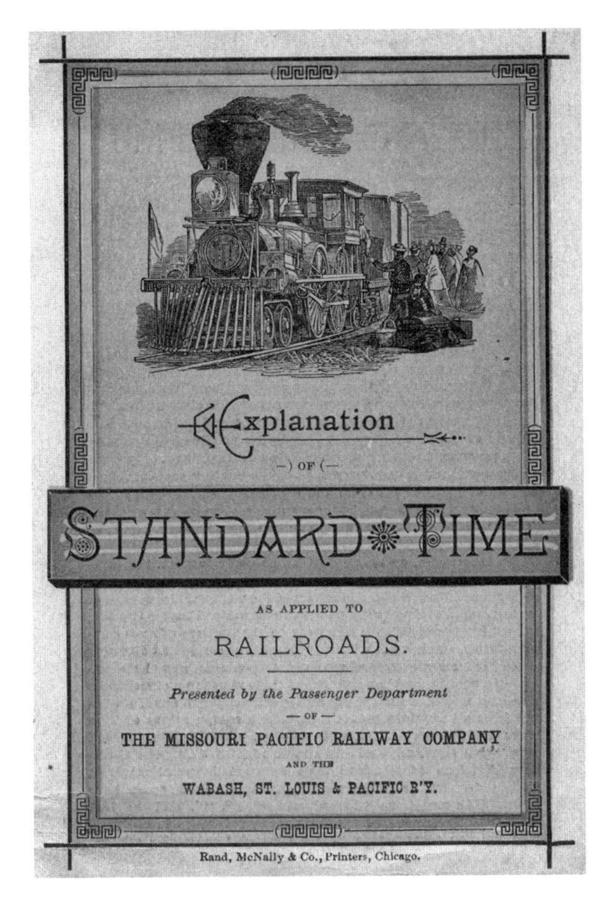

Both railroad employees and passengers needed to be educated about how Standard Time would be implemented by railroads across the nation in the fall of 1883. Railroads such as the Missouri Pacific published guides that explained the four new regional time zones in the United States and highlighted key differences between civil time and railroad time. *The Schwantes-Greever-Nolan Travel and Transportation Ephemera Collection, St. Louis, MO*

agreed to discuss the matter in detail at their next meeting.[36] In the spring of 1883, GTC members considered Allen's voluminous *Report on the Adoption of Standard Time* and unanimously authorized the secretary to contact member companies and "endeavor to secure [their] acquiescence . . . to the plan proposed."[37] At the next GTC meeting on October 11, 1883, a majority of GTC representatives resolved to observe the new time standards beginning just over a month later on November 18, 1883. By the end of the year, railroad companies nationwide were operating in accordance with the new time standards.[38]

The GTC's decisive actions regarding standard time encouraged members to tackle other operational concerns, most notably the lack of industry-wide uniformity in operating rules and train dispatching protocols. In addition to dealing with Allen's report on standard time at the GTC's spring meeting in 1883, the membership also considered a proposal to standardize the industry's use of visual and audio signals. James McCrea, a longtime Pennsylvania Railroad official, and the current manager of a number of PRR affiliated lines, convinced members to organize a new committee to research and develop standards for trackside, train-crew-operated signals.[39] In the fall of 1883, he presented the committee's findings. They proposed uniform requirements for color, flag, and whistle signals that were identical to or closely resembled the operating practices enumerated in the Pennsylvania Railroad's 1882 rule book. This was hardly surprising, given McCrea's involvement in formulating these rules as a member of the ATO. Secretary Allen was ordered to distribute copies of the new rules "to the managing officers of all the [rail]roads in the country, with the request that they will signify their assent or objection to the same."[40] A majority of respondents supported the proposed standards. General Superintendent Oliver S. Lyford of the Chicago & Eastern Illinois Railroad perhaps best summarized the general feelings of many officials when he replied, "I consider it not nearly as important what kind of signal is used as to have uniform signals on all [rail]roads."[41]

GTC members debated the new standards over the next year. At the spring 1884 meeting, a majority of attendees once again voiced their support for operational uniformity, but some members vociferously objected to completely altering their current operating practices to conform to the new PRR-based standards. Members also expressed concerns about the costs of implementing new signaling protocols on their lines and feared that train crews and station agents might make inadvertent operating errors as they transitioned to the new signaling practices. After hearing these complaints, committee member Edward B. Thomas of the Cleveland, Columbus, Cincinnati & Indianapolis Railroad scoffed, "If all the [rail]roads refuse to adopt the signals simply because they are those of another [rail]road, it will be impossible to secure uniformity."[42]

Some GTC officials also worried that the association was pressuring members to implement the new rules too quickly. Phineas P. Wright, GTC chairman and general superintendent of the Lake Shore & Michigan Southern Railway, a direct competitor of the PRR, argued that the Lake Shore could not possibly change its operating practices on short notice. Wright did not like the new PRR-inspired regulations and preferred to maintain the Lake Shore's current signaling code, which was based on the New York Central Railroad's operating rules. He complained, "We cannot expect that all [rail]roads will go to one [firm] and adopt its code." Wright lamented that if he were pressed on the issue, he would vote against the proposed changes.[43]

McCrea continued to push for a vote by GTC members. He stressed the importance of unanimity but also recognized that the convention was a voluntary organization and members could choose to ignore the GTC's proposed changes. He warned those in attendance, however, that Representative William Henry Calkins of Indiana was considering federal legislation on standard operating rules. McCrea noted that he had asked Calkins "to defer action for a while," so railroad officials could address the issue, but warned that time was running out. With the implied threat of federal regulation looming over them, the majority of GTC members resolved to approach their corporate directors and recommend adoption of McCrea's new standards in the fall of 1884.[44]

The GTC's hasty decision to adopt the PRR-based standards did not sit well with some industry observers. Correspondent Joseph E. Ralph complained in the *Railway Age* that many companies would not be able to safely implement the new rules in less than six months. He noted that even the PRR had taken nearly two years to implement some of the provisions of its 1882 rule book. Ralph argued that railroad officials should not rush to adopt new standards. They should spend as much time as necessary to develop clear and straightforward signaling practices that companies could safely and effectively use on their lines without disrupting their current operations.[45]

Despite these concerns, many firms belonging to the GTC voluntarily and speedily adopted the new standards, as they had standard time zones. In the fall of 1884, McCrea proudly reported that 116 firms had accepted the uniform signaling rules. On the other hand, thirty railroads had rejected the changes outright, and ninety-seven companies had indicated that they would consider the new measures only if connecting lines adopted them first. McCrea continued to pressure reluctant companies by warning that Representative Calkins had finally introduced a bill calling for uniform train rules. Based on this action, McCrea once again pressured recalcitrant members: "We trust that those [rail]roads which have not agreed to adopt the signals recommended by [the Uniform Train Signals Committee] will reconsider their decision."[46]

The decision by 116 railroads to accept uniform signaling standards represented an important initial step toward reforming operational management practices and promoting interfirm cooperation across the nation. While GTC officials publicly characterized the new standards as the product of interfirm cooperation and consensus building by GTC members, it is clear that McCrea used the implied threat of federal regulation to coerce industry participants into accepting the PRR-based signaling practices with little debate or modification.

McCrea's accomplishments as head of the Uniform Train Signals Committee may have motivated Superintendent Kirtland H. Wade of the Wabash Railroad to propose a new GTC committee for developing "Uniform Telegraph Orders and General Rules for governing train service." As a former telegrapher and train dispatcher, Wade recognized the problems that inconsistent dispatching and operating practices posed for the nation's growing rail networks. He wanted to develop a set of efficient and cost-effective dispatching rules that would permit railroad officials to safely manage train movements across different rail networks and make the best use of their limited telegraph resources. GTC members supported Wade's proposal, and he was asked to head the new Uniform Train Rules Committee.[47]

Like McCrea's Uniform Train Signaling Committee, Wade's new committee contained influential officials from across the Pennsylvania Railroad. Train Rules Committee member Robert Pitcairn had served alongside McCrea as a member of the PRR's Association of the Transportation Officers and had been involved in developing dispatching and block-signaling rules for the PRR's 1882 rule book. Wade also invited PRR superintendent and telegraph dispatching expert John A. Anderson to participate in committee meetings as a guest and share his extensive knowledge of dispatching practices with the group.[48]

Due to the scale of the project, more than a year passed before Wade's committee presented the first draft of the proposed universal rule book to the general GTC membership.[49] At the fall 1886 meeting, GTC officials spent two days reviewing every proposed rule. Many members voiced concerns about the language used in the new rule book, and some found the rules far too specific. Former Erie Railroad dispatcher Charles Darius Hammond, who represented the interests of the Delaware and Hudson Canal Company, complained that committee members were making the new code much too detailed. He argued that complicated, PRR-based operating and dispatching rules would breed confusion among employees working for smaller railroads and constrain the decision-making abilities of divisional superintendents.[50] Charleston & Savannah Railroad general manager Henry Stevens Haines worried that many railroad employees would be overwhelmed by the complexity of the new rules and fail to follow them. Despite the concerns expressed by Hammond, Haines, and other officials, a majority of GTC members voted to provisionally accept the new operating rules until Wade's committee worked out the final details.[51]

The following spring, Wade presented the revised code. He acknowledged Anderson's assistance and thanked him for sharing with the committee "the benefit of his long and well-known experience in the formation of Telegraphic Train Rules."[52] Not surprisingly, the code reflected Anderson's general outlook on train dispatching and shared many similarities with the PRR's 1882 rule book. GTC members received many of the proposed operating rules positively, but the greatest controversy arose from the committee's decision to adopt Anderson's double-order train dispatching protocol. Some members complained that the greater volume of telegraph messages sent under the double-order dispatching method would tie up their limited telegraph resources and put unnecessary stress on dispatchers and train crews. These members asserted that specific dispatching practices should be left up to the discretion of individual companies, not GTC mandate. Uniform Train Rules Committee member Henry B. Stone, general manager of the Chicago, Burlington & Quincy Railroad, disputed these arguments. Stone asserted that double-order dispatching would prove safer and more reliable than the single-order practices then in use on many railroads. He noted that double-order dispatching would promote uniformity by ensuring that all train crews receive identical train orders. This would prevent dispatching errors that might lead to costly collisions.[53]

GTC members considered both perspectives and voted in favor of the new, uniform operating rules. Over the next two years, GTC members continued to refine the new dispatching requirements at biannual meetings. Some members remained concerned about the costs of implementing the new operating standards and argued, in particular, that double-order dispatching was labor intensive and prohibitively expensive on their lines. Pitcairn countered that the double-order dispatching was far more efficient than older practices and would allow poorly capitalized railroads to make better use of their limited telegraph resources by sending fewer train orders each day, though the messages would be longer since they would include all the current orders rather than a single order for one train crew. He argued that his fellow committee members could "cite instance upon instance where, under the single-order system, no single and no double telegraph wire would enable you to send all the orders which are successfully, and for years have been successfully, transmitted over a single wire by the duplicate [double]-order system."[54]

Despite these assurances, a few members remained unconvinced by Pitcairn's arguments. Hammond of the Delaware and Hudson Canal Company argued that the new rules would not "decrease or in any way lessen the dangers attendant upon the moving of trains by special telegraph orders." They were simply "a great deal of burdensome lumber" for his small railroad. He concluded: "I have, I think, but eight or ten rules, covering everything relating to the moving of trains by special orders. I have seen no reason, after fifteen years of working under these rules, to change a single one of them."[55]

Hammond, like other managers of lightly trafficked, small rail lines, remained deeply concerned about the risks of dispatching trains by telegraph. He failed to see how new, uniform rules would lessen the inherent dangers associated with the practice. Rules could not prevent dispatchers, telegraphers, or other members of the dispatching chain from making mistakes, which could lead to horrific accidents. Unlike Pitcairn and other managers employed by the vast, modern PRR network, Hammond had not accepted telegraphy as a normal part of daily operations. His comments about moving trains by "special orders" indicated that he still considered the practice exceptional, not routine. According to his viewpoint, managers should minimize their dependence on telegraphy, instead of attempting to develop new and sophisticated rules to regulate an inherently dangerous means of communication.

The GTC membership discounted Hammond's specific concerns about telegraphy and overwhelmingly voted to add the new, uniform dispatching practices to the Convention's Standard Code of Train Rules. A year later, the members of the new Committee on the Standard Code of Train Rules and Rules for the Movement of Trains by Telegraphic Orders proudly reported that forty-five firms had adopted the Standard Code with minor revision, and an additional twenty-three railroads planned to do so in the near future. By April of 1890, ninety-three companies operating over sixty-five thousand miles of track reported that they were using the Standard Code in their daily operations.[56]

In 1893 when the Railroad Gazette Publishing Company reissued *The Train Wire*, Anderson's influential series of essays about telegraphic train dispatching originally published ten years earlier, railroad signaling expert Braman B. Adams noted in his introduction that Anderson's views on double-order dispatching had become "widely recognized as an axiom" on most railroads thanks to his work with the GTC's Train Rules Committee.[57] Anderson's guiding influence over fellow PRR superintendents James McCrea and Robert Pitcairn, as well as other GTC officials, had helped produce consistent and universal operating and train dispatching standards. These officials had successfully argued that modern railroad firms required new managerial tools such as telegraphy to operate safely and efficiently. As Adams concluded in his introductory remarks, "The introduction of the Standard Code on 70,000 miles of American railroads is one of the important steps of recent years in railroad operation, resulting in greater security to life and property." He further noted that Anderson and *The Train Wire* "should be credited with a liberal share of the honor of the reform."[58]

Conclusion

Undoubtedly, the General Time Convention's Standard Code reflected the Pennsylvania Railroad's particular operating standards rather than a true consensus among railroad officials about the best operating protocols for the nation's many railroads. The code emphasized the importance of modern and cutting-edge operational practices used by a few wealthy and powerful railroads in the United States, rather than sanctioning older and more traditional methods used by many of the nation's smaller and less-capitalized railroads. Despite the vocal opposition of managers such as Superintendent Hammond and others who expressed concerns about the financial and managerial implications of the new operating and dispatching requirements, PRR officials succeeded in convincing the GTC's membership that PRR standards represented the best approach for promoting safety and efficiency on the nation's many miles of railroad track, particularly with respect to overseeing train movements via the telegraph.

Ironically, American railroad officials by this era had access to a multitude of equipment, both mechanical and electrical, that could reduce or even eliminate the dangers associated with operating trains on their rail lines. Block signaling and various centralized train control methods held the potential to safely regulate train movements with minimal human input. Yet, as the GTC's deliberations throughout the 1890s and the early years of the twentieth century reveal, officials continued to prioritize organizational solutions over expensive technical fixes, especially when it came to dealing with operational management problems. Despite concerns about the growing army of dispatchers and telegraphers who controlled train movements on their lines, managers were unwilling to consider adopting electrical communication or signaling devices that could effectively replace many of these members of the dispatching chain or reduce their influence over rail operations. Ultimately, political, regulatory, and labor pressures in the first decade of the twentieth century would force railroad officials to confront their fears about the disruptive influence of new communication devices on railroad operations and bring about significant change throughout the American railroad industry.[59]

Telegraphers and Regulators

> Under the present system the [train] engineer and the train dispatcher take the responsibility and we have to submit to their liability to make mistakes: is the telegraph operator anything but one more fallible guardian?
>
> "General Manager," *Railroad Gazette*, 1893

Throughout the 1890s, a growing number of American railroad officials adopted the General Time Convention's Standard Code as the basis for their firms' operational management practices. The newfound emphasis on standardization, particularly with respect to telegraphic train dispatching and signaling practices, reduced the number of operational inconsistencies across different railroad lines that had contributed to numerous train collisions throughout the 1870s and 1880s. Standardizing operating rules and train control practices, however, addressed only half of the managerial challenges faced by officials at the end of the nineteenth century. Standardized rules and regulations could not resolve growing labor conflict between railroad managers and the vast cohort of telegraphers who were responsible for moving trains safely and efficiently across America's expanding rail network. Officials routinely characterized telegraphers as flighty, troublesome, and at times even irresponsible, but they were utterly dependent on these "fallible guardians" to keep railroad traffic moving day and night.[1] Many telegraphers, cognizant of their central place within the railroad management hierarchy, resented the poor pay, long hours, and difficult working conditions that they faced in railroad service. In the late 1880s, disgruntled midwestern railroad telegraphers organized the Order of Railroad Telegraphers (ORT) to fight for better pay and working conditions, and the new railroad brotherhood spread nationwide over the next two decades. At first, railroad officials resisted ORT demands for higher wages and shorter working hours, but over time the ORT grew in strength and formed alliances with other railroad brotherhoods and with progressive political reformers who had begun to gain power and influence at the state and national levels around the turn of the twentieth century. This new coalition of labor organizers and political reformers had a profound impact on the railroad industry during the first decade of the twentieth century.

The alliance between the railroad brotherhoods and progressive politicians produced several important pieces of reform legislation. Of these, the 1907 Hours of Service Act ("HOSA") had the most significant impact on railroad

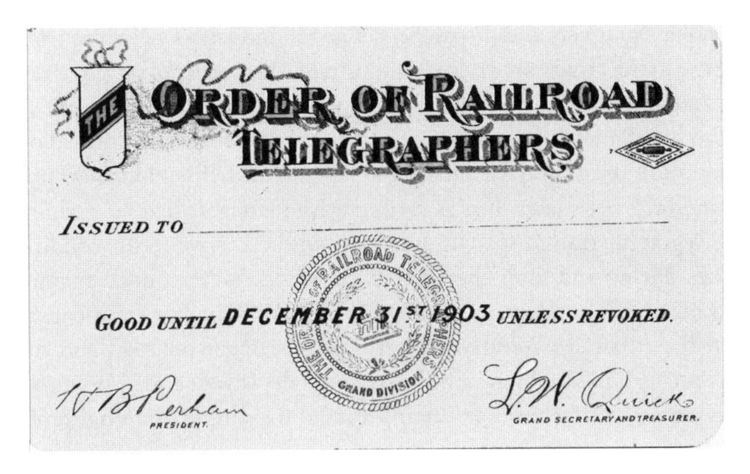

Founded in 1886 as the Order of Railway Telegraphers of North America, the protective association changed its name to the Order of Railroad Telegraphers in the early 1890s. The organization emerged from an earlier telegraphers' union that had been part of the Knights of Labor. Members, however, gradually came to the realization that railroad telegraphers' work was significantly different from the work performed by commercial telegraphers employed by Western Union and other telegraph firms of the era. By the beginning of the twentieth century, the union had more than 10,000 card-carrying members and had been recognized by railroads across the nation. *The Schwantes-Greever-Nolan Travel and Transportation Ephemera Collection, St. Louis, MO*

operations during the era. The law, which went into effect in March 1908, strictly limited the number of hours that railroad employees, including telegraph operators and train dispatchers, could work in a twenty-four-hour period. The act effectively forced American railroads to double their workforce of telegraphers in less than a year, an impossible task by 1907 due to the growing strength of the ORT and their success in limiting the number of new railroad telegraphers learning the trade. In desperation, railroad officials across the country began hastily to replace or supplement their extensive telegraphic train dispatching networks with a new communication device, the telephone. Despite profound reservations, officials with the Chicago, Burlington & Quincy (CB&Q), the Illinois Central (IC), the Southern Pacific (SP), the Pennsylvania (PRR), and other busy, main-line railroads shifted a considerable volume of train dispatching from telegraph to telephone circuits between 1907 and 1910.

Following the switch, railroad officials throughout the country began to praise the telephone as an ideal tool for operational management. Some even argued that it was better suited for managing railroad operations than the telegraph and its army of telegraphers and dispatchers. The rail industry's switch to telephones in response to the Hours of Service Act illustrates the impact of progressive reform politics and unionization on railroad management in the early twentieth century, but it also highlights important similarities between American railroad officials' rapid adoption of telegraphy during the 1860s and their sudden acceptance of telephony forty years later. Just as conservative managers had begrudgingly accepted telegraphic train dispatching to cope with excessive traffic conditions on their lines during and after the Civil War, outside factors, once again, forced railroad managers to adopt a new device, the telephone, for train control purposes in the twentieth century. In both cases, exigent political and financial circumstances forced reluctant officials to act decisively and apply innovative technical solutions, rather than organizational fixes, to operational problems.[2]

Troublesome Telegraphers

By the 1890s, American railroad officials depended on nearly twenty thousand telegraphers and train dispatchers to maintain safe and efficient operating conditions across more than 150,000 miles of rail line.[3] However, many in the industry were gravely concerned about the competence and reliability of this army of employees who served as the crucial link between divisional headquarters and individual train crews. No matter how many safeguards superintendents built into railroad rule books, the "fallible guardians" continued to make mistakes that contributed to accidents and deaths. In 1899, eighteen people were killed in a collision on the Lehigh Valley Railroad after a telegraph operator neglected to write down a dispatcher's instructions before communicating them inaccurately to a train crew. The Lehigh had adopted the General Time Convention's Standard Code in 1894, but the rigid rules had failed to prevent or mitigate the employee's error. Not surprisingly, the Lehigh Valley crash, and similar incidents across the nation, led industry observers to identify telegraphers as a weak link in the increasingly orderly and efficient American system of train dispatching.[4]

Telegraphers proved to be a constant source of frustration for railroad managers during the 1890s and early 1900s. Many officials considered them to be nearly unmanageable workers. In an industry in which high labor turnover within certain positions was the norm, telegraphers had some of the shortest employment rates of any employees. A survey of sixty major railroads in 1889 determined that the average period of employment for telegraph operators was approximately five months. The itinerant, "boomer" telegrapher, constantly in search of adventure and better wages, was a stock romantic character in popular railroad magazines of the period. In practice, though, such a high turnover rate meant that divisional superintendents were constantly searching for telegraphers to fill open posts along their rail lines and were willing to hire less-than-competent operators, if necessary, to staff certain remote stations.[5]

Railroad officials bore a great deal of the blame for this unreliable and footloose telegraphic workforce. Railroads chronically overworked and underpaid telegraphers. To save money, managers often hired young, inexperienced telegraph operators and required them to work long hours in remote stations.[6] Officials economized by assigning station operators additional duties. Telegraphers might be guaranteed a daily wage for their services, but they would also have to send commercial messages for Western Union, serve as station postmasters, and perform other time-consuming acts. As a result, Western Union and the federal government indirectly subsidized station operators' wages, and the railroads paid only a small portion of their overall earnings. The additional work wore down operators physically and mentally during the course of their twelve- to sixteen-hour, seven-days-a-week shifts and bred resentment against their parsimonious employers.[7]

Beyond simply representing a labor management problem, overworked and inexperienced telegraphers represented a clear and present danger to railroad operations. Sleeping operators failed to alert dispatchers as trains passed their stations. They neglected to set stop signals to halt trains to pass along new train orders to crews. Overworked operators overlooked errors in train orders from dispatchers. Inexperienced operators made bad judgment calls instead of contacting superiors for guidance. All these situations increased the likelihood of deadly collisions, but many superintendents accepted the risk because they believed that their firms' operating rules and regulations provided sufficient safeguards against disaster.[8]

Veteran railroad telegraphers made no secret of their frustration with poor pay, long hours, and inexperienced coworkers. In 1892, correspondent "Ajax" wrote to the *Railroad Gazette* about the aggravating working conditions that telegraphers faced nationwide. "Small and unreasonable salaries" drove older and more experienced operators with families away from railroad service and forced younger, less experienced telegraphers to adopt the itinerant boomer lifestyle to make a living. The declining quality of operators and high turnover rate lowered the reputation of all railroad telegraphers in Ajax's opinion. He complained that "the fact that you are an operator is sufficient cause for being looked down upon" by other railroad employees and managers. He argued that railroad officials must pay competitive wages and promote talented operators to positions of managerial authority if they hoped to retain qualified telegraphers on their rail lines.[9]

The editor of the *Railroad Gazette* sympathized with Ajax's complaints about the conditions faced by railroad telegraphers, agreeing that managers often treated operators poorly with respect to wages, hours, and working conditions. However, he asserted that the true cause of telegraphers' poor wages and long hours was the crowded labor market of the early 1890s. "There are two qualified applicants for one place where there were formerly two places for each good man," the editor reminded readers. No longer could telegraphers command the "high wages incident to a new industry," as they had when railroads had first begun hiring operators in the 1850s and 1860s. Similarly, reform-minded telegraphers would have no success obtaining shorter working hours from railroad officials "as long as there are plenty of men willing to work 15 hours for the same pay" as someone working ten hours. Instead, the editor argued, telegraph operators must be more realistic regarding pay and hours of labor.[10]

While the *Gazette*'s editor advised telegraphers to temper their expectations, he also had a warning for railroad officials who thought that they could continue to pay operators low wages in exchange for lengthy working hours without any long-term repercussions. The editor cautioned that such behavior would encourage unionization among railroad telegraphers and eventually lead to regulatory pressure at the state and federal levels. "It is only to be expected that the overworked men will enlist the sympathy of legislators and of the public generally," he noted. Railroad officials, the editor warned, must avert unionization and adverse regulatory legislation by treating their growing workforce of operators with greater respect or face unpleasant consequences.[11]

By 1892, perceptive industry observers such as the *Gazette*'s editor had good reason to be concerned about unionization among railroad telegraphers. Efforts to develop a national brotherhood for railroad telegraphers had been underway for nearly six years. In 1886, railroad telegraphers from the Midwest had organized the Order of Railway Telegraphers of North America, later changed to the Order of Railroad Telegraphers (ORT), to represent the interests of operators engaged in railroad businesses. Though originally created as a fraternal association, by the early 1890s the ORT was transformed by new leadership into a protective union, and membership grew to more than nine thousand operators. Union leaders argued that higher wages and shorter hours would create a more stable workforce of skilled railroad telegraphers and ultimately benefit railroads by reducing the number of inexperienced boomers in railroad service. The ORT leadership also sought to restrict the labor supply of new telegraphers entering railroad service. The order's constitution required members to seek the personal approval of the ORT president before they could take on new apprentices. While union leaders later admitted that the provision, in and of itself, did little to dramatically reduce the number of telegraphers available for railroad work during the 1890s and early 1900s, it did reflect the ORT's awareness that an overabundance of telegraph operators eroded their bargaining position with railroad firms.[12]

The ORT enjoyed mixed success during the 1890s. Members struck unsuccessfully against the Milwaukee & St. Paul Railroad in 1891 and the Rock Island Railroad a year later, but union organizers remained persistent and eventually secured recognition by management on approximately twenty-five major railroads across the country before the Panic of 1893 led railroad officials to take a harder line against the union. ORT membership declined as a result of the panic and remained stagnant until late in the decade, but the business contraction forced the union to become more organized and seek strategic affiliations with other major railroad unions of the era, including the Brotherhood of Locomotive Engineers, the Order of Railway Conductors, and the Brotherhood of Railroad Trainmen. As railroad business improved after 1897, the ORT resumed its efforts to secure recognition by railroad management. By 1901, the order claimed more than ten thousand active members, and the number quickly grew to thirty-seven thousand by 1907.[13]

The growing strength of the ORT in the early 1900s, combined with the dramatic reorganization and expansion of American railroad networks as the nation recovered from the Panic of 1893, renewed managerial concerns about the mounting shortage of skilled, nonunion telegraph operators in railroad service. In 1904, J. C. Brown, the general foreman of the St. Louis, Iron Mountain & Southern Railway, took fellow railroad officials to task in print for employing less-than-competent station telegraph operators and failing to educate and train a new generation of telegraphers in-house. Brown noted that "frequent merging of small railroads into large railroad systems during the past ten years has increased the demand for improvements in the railroad telegraph department," but the writer complained that the "general qualification of railroad operators is far below what it was ten years ago." Commenting on Brown's article, the editor of the *Railroad Gazette* worried about another issue caused by the decline in qualified operators. He argued that the decline had forced railroad officials to become more reliant on individual train dispatchers to "safeguard against errors by a dozen incompetent station-operators." In the long term, the *Gazette*'s editor noted, the practice had placed "an unreasonable burden on the dispatcher," and "the safeguard often breaks down." Overworked dispatchers made many of the same errors as overworked telegraphers, which led to collisions that fostered public concerns about the safety of American railroads.[14]

Railroad officials' growing reliance on train dispatchers to keep the American system of train dispatching running smoothly despite the growing presence of young and inexperienced telegraphers created another potential source of labor conflict. Dispatchers viewed themselves as managers in their own right, since they issued orders to station telegraphers from divisional headquarters and made crucial decisions on their own about train movements on rail lines under their oversight. In the late nineteenth century, the work performed by dispatchers was often publicly depicted in heroic terms. In 1882, the *Boston Herald* described the consequences of a minor accident on the Old Colony Railroad of Massachusetts. The writer noted that the derailment of a freight train immediately brought fifty-five trains to a standstill on the busy commuter line. To clear the line, the train dispatcher, the "king of the situation all the day long," faced the daunting challenge of rapidly issuing orders to dozens of delayed trains. The author emphasized the dispatcher's taxing responsibilities: "A minute too soon or too late with this or that train; a blundering order, or one not quite definite; the mistake of a minute or a mile—how many contingencies hang upon his knowledge and action!" Such comments portrayed dispatchers as powerful, yet potentially flawed, supervisors with grave responsibilities constantly weighing on their shoulders.[15]

Public comments by dispatchers further emphasized the demanding nature of their work. A dispatcher writing to the *Cleveland Herald* described the physical tolls of his job: "The train dispatcher has a hundred trains under his finger that pressed upon a [telegraph] key. . . . The eight hours you put in bending over your keys seems like a week. Your head swims and grows dizzy beneath its awful responsibility."[16] Other newspaper articles emphasized the unique mental character of a good dispatcher: "A dispatcher must have prudence, a wonderful memory, a very accurate mind and great power to act promptly on information received. . . . He brings his judgment to bear on all these points as he gives his orders, and his ability is judged first of all by his record for safety and secondly by his record for ingenuity in keeping all trains moving steadily forward."[17] Dispatchers represented the heart of railroad operations, as the articles illustrated, but their duties wore them down physically and mentally and made them prone to committing errors that could endanger the traveling public.

Like telegraphers, dispatchers began to complain that railroads were failing to properly compensate them for their contributions to railroad safety and efficiency. Their protests grew more fervent during the 1890s. Dispatchers had organized the Train Dispatchers' Association in the mid-1880s as a fraternal organization, and at the time they had eschewed all questions of unionization and instead pledged their allegiance to their employers. In the early 1890s, however, association members became more vocal in their demands for shorter hours and higher wages. In 1892, John F. Mackie, chairman of the executive committee of the Dispatchers' Association, declared that many dispatchers believed that railroad companies were treating them poorly. He felt, however, that dispatchers would continue to remain loyal to their companies if superintendents would address these inequities.[18]

Mackie's comments generated a heated debate among dispatchers. Some complained that managers treated them no better than itinerant, nonunion telegraphers. They argued that dispatchers should defect from the Dispatchers' Association and join the ORT to fight for better hours and wages. Others argued that they had more in common with white-collar managers and fought efforts by the ORT to recruit dispatchers into the union. In an 1894 issue of *Railway Age*, Dispatchers' Association officials offered a list of nearly two hundred senior railroad officers who had served as dispatchers at various points in their careers. They cautioned fellow dispatchers not to "put any stumbling blocks in their way" by joining unions. Instead, they argued, dispatchers could earn promotions to managerial positions only through "earnest application, mastery of their work and devotion to duty." Despite these warnings, some dispatchers abandoned the association and adopted a more confrontational attitude toward their employers.[19]

The growing concerns over the dependability of overworked and possibly disgruntled telegraphers and dispatchers spilled into the public arena. Like managers, the general public was aware of the grave consequences that could arise because of railroad employee errors. Newspaper articles from the late nineteenth century hinted at some of these dangers to the traveling public. In "His Life Ruined by a Cipher," an 1890 short story in the *San Francisco Examiner*, a luckless (and possibly overworked) train dispatcher makes a simple error that sends two passenger trains on a collision course. Realizing the error a few minutes later, "Bill" can do nothing to stop the inevitable accident. He telegraphs the head office for a repair train to clean up the wreckage and submits his resignation while waiting for news of the collision. The author of the story tapped into popular anxieties about railroad safety and highlighted the potentially unreliable character of employees in positions of authority. He noted that a dispatcher's omission of a single digit from his train order, while seemingly a minor mistake, could mean the difference between life and death for train passengers in the dangerous world of railroad operations.[20]

For early twentieth-century railroad officials and the public, dispatchers and telegraphers represented potentially unreliable and unsafe components in an increasingly sophisticated and uniform railway management framework. While officials had made great strides to develop and implement industry-wide operating rules and dispatching practices in the final decades of the nineteenth century, they depended on overworked and underpaid employees to safely execute the new uniform operational rules. Officials had made telegraphic dispatching a central component in railroad operations, but they could not remove the human element. Furthermore, their efforts to save money by hiring less-than-competent operators and overworking them only created greater problems for managers and increasingly exposed the industry to criticism from the public and reform-minded politicians. In the first decade of the twentieth century, growing public concerns about railroad safety and the dangers posed by overworked employees would lead to significant federal legislative reform efforts that would have a direct impact on railroad management.

Hours of Service

The final decades of the nineteenth century witnessed growing public concern over real and perceived public safety risks on American railroads. For both railroad employees and the traveling public, American railroads seemed no safer than they had in the post–Civil War period. During the 1880s and 1890s, industry officials responded to public fears and growing pressure from state and federal legislators by implementing internal reform campaigns such as the General Time Convention's efforts to develop the Standard Code and the Master Car-Builders' Association's work on standardizing rail car parts such a couplers, axles, and wheels. Such efforts did not entirely stave off regulatory legislation. In 1887, Congress passed the Interstate Commerce Act and empowered the Interstate Commerce Commission (ICC) to regulate rates charged by railroads for interstate shipping. Over the next six years, reformers pressured Congress to expand the ICC's purview to include railroad safety. Petition drives and public awareness campaigns by railroad brotherhoods attracted public attention to the dangers faced by brakemen as they attempted to couple cars and operate hand brakes on moving trains. Railroad officials initially opposed congressional efforts to regulate railroad safety, but they eventually recognized the futility of their efforts and instead sought to shape the new legislation in ways that would not harm the industry.

Congress passed the Safety Appliance Act in 1893 with the stated purpose of promoting "the safety of travelers and employees on the railroads."[21] The act stipulated that all trains must be equipped with braking equipment that would allow the locomotive engineer to control the speed of the train without having to depend on brakemen to tighten hand brakes on individual cars. The act also directed the American Railway Association, the national industry association that grew out of the General Time Convention, to develop standards for coupler design and location on cars to prevent injuries and accidents resulting from incompatible equipment. For the most part the new law "embraced current practice by the best firms and helped railroads ensure standardization and a level playing field."[22] Congress placed the ICC in charge of implementing the new law and ensuring compliance, but as a result of various extensions, it did not go into full force until the early twentieth century.

The 1893 Safety Appliance Act, and a 1903 amendment that addressed shortcomings in the original act, focused on reforming and regulating railroad equipment as a means for promoting the safety of train crews and travelers. During the first decade of the twentieth century, however, progressive reformers and industry regulators at the national level began to draw attention to a new threat to the traveling public. In his fourth annual message to Congress in December 1904, President Theodore Roosevelt lobbied members on both sides of the aisle to regulate the working hours of railroad employees engaged in interstate commerce and ensure that railroads would hire only qualified and experienced individuals to operate trains and oversee train movements. A few weeks later, the Interstate Commerce Commission released its annual report in which commissioners expressed concerns that excessive working hours might be a major contributory factor in numerous railroad accidents. The commissioners called for an investigation of the subject to determine the dangers that overworked railroad employees posed to the traveling public.[23]

Despite the White House's and the ICC's efforts to raise awareness of the subject, Congress did not act. A year later, Roosevelt repeated his demand for congressional action on the matter of railroad employees' working hours. The president argued that railroad employees experienced more stress than perhaps any other occupation and depended on "vigilance and alertness" to perform their duties. He argued that Congress must act to protect the safety of travelers on America's railroads. Similarly, the ICC's 1905 annual report released shortly after Roosevelt's address once again highlighted the public danger posed by overworked railroad employees and called for congressional action on the matter.[24]

Congress finally began to examine the issue of railroad employees' work hours in the spring of 1906. The House Committee on Interstate and Foreign Commerce convened a series of hearings on the matter and took testimony from industry officials and representatives of the railroad brotherhoods. Officials expressed concerns that legislation limiting the hours of railroad employees directly involved with train moments would disrupt rail service by necessitating crew changes for locomotive engineers, firemen, and brakemen at inconvenient times and locations. Railroads would also be forced to hire additional station telegraphers and dispatchers to maintain train dispatching services around the clock. In all, rail officials argued, any restrictions on railroad employee hours would be prohibitively expensive and difficult to implement nationwide.[25]

In May of 1906, two bills appeared before Congress concerning hours of service for railroad employees. Following weeks of hearings, the House Committee on Interstate and Foreign Commerce reported a bill authored by Wisconsin Republican John Esch to the floor of the House. Esch's bill stipulated that railroad employees on trains must receive a ten-hour rest break after working sixteen hours, and telegraphers and train dispatchers could work only nine consecutive hours in a twenty-four-hour period. Railroad officials would be fined if employees violated these stipulations. Observers considered the Esch bill pro-railroad in nature because it contained a number of loopholes that would make it difficult for the ICC to pursue violations. Officials would receive fines only if they "knowingly" allowed employees to work beyond the stipulated time limits or forced them to do so. The bill contained no enforcement mechanism for assessing fines and did not permit the ICC to investigate railroad compliance with the law, only to respond to specific complaints.[26]

Across the Capitol building, the Senate considered a very different railroad bill. Senator Robert La Follette, acting at the behest of Theodore Roosevelt, introduced his own hours of service bill. In some respects, La Follette's bill resembled Esch's bill. Railroad employees operating trains could work only sixteen hours continuously before they had to take a ten-hour break. Telegraph operators and dispatchers were restricted to nine hours of work in a twenty-four-hour period. Unlike Esch's bill, however, La Follette's measure did not contain any loopholes that would allow railroad officials to use ignorance of violations as an excuse to avoid penalties. Whether or not officials knew of or condoned overwork, they would be held accountable under La Follette's legislation. The Senate bill also gave the ICC full authority to investigate railroad compliance with the act and pursue violators in federal court if necessary.[27]

Neither the full House of Representatives nor the full Senate chose to pursue their respective railroad hours-of-service bills during the 1906 session. La Follette pressured the Senate to consider his bill, which sparked a heated debate on the Senate floor over the measure, but he could not compel a vote. Instead, the Senate leaders—anxious to begin their summer recess and concerned about voting on the controversial bill prior to the fall election—agreed to postpone any vote on La Follette's hours-of-service act until early January 1907.

The postponement gave the railroad industry ample time to lobby against La Follette's and Esch's proposed bills. In December, the editor of the *Railway Age* condemned the bills as unwelcome government paternalism that would have a detrimental impact on railroad firms and employees alike. Railroads would face serious disruptions in their schedules during busy traffic periods, as they would not have enough employees to run trains around the clock and keep telegraph offices and divisional train dispatching centers staffed. Employees would suffer, since they would lose opportunities to earn additional wages by working extended hours. To support his argument, the editor of the *Railway Age* cited a resolution recently passed by Division 156 of the Order of Railway Conductors that condemned the La Follette–Esch bills for interfering with employees' freedom of contract with their employers and their right to work as they saw fit. The trade journal's editor viewed this resolution as a sign that railroad employees did not support the protective legislation any more than railroad officials.[28]

Robert La Follette questioned the legitimacy of employee resolutions against the pending legislation. He asserted that railroad officials had pressured local railway lodges such as Division 156 into publicly condemning the Esch and La Follette hours-of-service acts. He cited a case in which a divisional superintendent for the Southern Railroad had presented a local employee lodge with a prepared statement critiquing the La Follette–Esch bills and encouraged the lodge, unsuccessfully, to back the resolution. La Follette argued that he had support for his legislation from railroad employee organizations in forty-three states and would not be deterred by railroad officials' efforts to generate false grassroots opposition.[29]

In January of 1907, the full Senate resumed debate on La Follette's hours-of-service bill and passed the measure by a mere two votes. The House Committee on Interstate Commerce next took up the measure in mid-February, and Republican William Hepburn of Iowa, the powerful chair of the committee who had been responsible for the 1906 Hepburn Act giving the ICC power to set railroad rates, substituted the weaker Esch bill in place of La Follette's legislation for consideration by the full House. Observers speculated that the substitution would ultimately kill the legislation and spare the railroad industry from La Follette's reforms. The *New York Time*'s reporter went so far as to call it a "snub of [Theodore] Roosevelt . . . [who had] been very active in his efforts to secure the passage of the bill . . . with a minimum of changes." The Esch substitution bill passed the House, but few observers expected it to become law.[30]

On March 1, a conference committee composed of members from both chambers met to reconcile the differences between the House and Senate bills. The committee recommended a reduction from ten hours to eight hours in the rest period that train crews must observe after working for sixteen hours continuously. They also increased from nine to twelve the number of hours that telegraph operators and train dispatchers could work in a twenty-four-hour period. Lobbyist H. R. Fuller, who represented the interests of the Brotherhood of Railroad Trainmen and a number of the other railroad unions, condemned the changes and appealed to railroad employees, particularly members of the Order of Railroad Telegraphers, to make their objections to the conference committee's actions heard throughout the halls of Congress.[31]

The following day, more than twenty thousand telegrams flooded both houses of Congress from railroad telegraphers across the nation. A *New York Times* reporter described the unprecedented scene:

> The railroad telegraph operators did not need to understand all the ins and outs of the work that was being done in the conference. It was sufficient for them to know that their bill was being bludgeoned to death. . . . Every operator had at hand the means of prompt communication with his Senators and Representatives. Before night messages began to arrive for Senators and Representatives protesting against the massacre of the bill. Today they come in floods. Every operator in the country sent at least one message to his Congressman and to each of his Senators. And then he went out and got his friends to telegraph also. . . . Nowhere on earth do men run to cover more promptly than in Congress when it becomes apparent to them that public opinion is aroused. The operators had succeeded perfectly in creating the impression that it was aroused. The result was that the members of both Houses labored with the conferees to put the bill in proper shape.[32]

Fuller's appeal to the ORT had its desired effect, and the sudden groundswell of grassroots opposition to the conference committee's amendments forced committee members to reevaluate their position.

On March 3, the committee announced the final amendments to the Hours of Service Act. The revised bill stipulated that telegraphers and dispatchers could work nine hours in a twenty-four-hour period in stations that were open both day and night and thirteen hours in stations that were open only during the day. The legislation took the uncertain nature of railroad operations into consideration by allowing telegraphers and dispatchers to work four additional hours each day in the event of an emergency, but the extended hours could not exceed three consecutive days in a week. The Interstate Commerce Commission was granted regulatory authority to enforce the provisions of the act and pursue violators in federal court for an amount "not to exceed $500 for each and every violation."[33]

Both houses of Congress approved the amended Hours of Service Act (HOSA) on March 4, 1907, and Theodore Roosevelt signed the measure the same day. The legislation gave railroad firms exactly one year to comply with the terms of the act and provided the ICC with narrow leeway to extend the compliance deadline for individual firms "in a particular case, and for good cause."[34]

The railroad trade press remained skeptical of the law following its passage. The editor of the *Railroad Gazette* thought it unlikely that a "bureau managed in Washington" would be able to monitor more than 200,000 miles of rail line stretching across the nation and felt that the new law did not address many of the hidden dangers that disrupted railroad operations and threatened the lives of passengers and crew. Still, the editor felt that HOSA might be useful as an "instrument for indirect coercion" of railroad officials who refused to adopt modern management practices and expressed faith that "wise railroad manager[s] will set [their] house in order long before next March."[35]

The passage of HOSA created a new regulatory function for the ICC and expanded its authority over railroad management and safety. The act also forced industry officials across the country to reevaluate how they utilized their operational workforce, particularly the army of telegraph operators and dispatchers who staffed thousands of large and small stations along nearly every rail line in the nation. Managers, who had struggled with shortages of telegraphers and dispatchers before HOSA passed, now had to contend with a law that required them to double their workforce of telegraphers to maintain the same level of train dispatching on their rail lines. The growing strength of the ORT, and the union's instrumental role in securing shorter hours for telegraphers and dispatchers, gave managers no reason to believe that the labor shortage would simply resolve itself on its own. Over the next few years, railroad officials pursued a variety of strategies to mitigate the labor shortage posed by HOSA.

Railroads Respond

During the year that followed the passage of the Hours of Service Act, the American railroad industry simultaneously sought to challenge or delay implementation of the law and pursued a variety of short-term organizational solutions to comply with the law's strict labor provisions. Despite the *Railroad Gazette*'s prediction that managers would set their houses in order before ICC enforcement of the Hours of Service Act began in March 1908, many railroad officials assumed something of a "wait and see" attitude toward the new legislation during the summer and fall of 1907. Only as the March 4 deadline grew closer did officials begin to put pressure on Congress and the ICC to postpone implementation of the law, particularly the nine-hour limit for telegraphers. In February 1908, more than fifty major railroads, including the CB&Q and the Southern, filed petitions with the ICC requesting exemption from the telegraph provision of HOSA. At the same time, the ORT continued its campaign to ensure that telegraphers would be covered under the new law. Members sent more than seventeen hundred telegrams to ICC commissioners calling on them to reject the railroads' appeals.[36]

The commissioners agreed to hold a hearing at the end of February to consider the railroads' petitions for individual extensions, but they made it clear that they would grant extensions only in the event of "exceptional instances of unusual conditions which [rail]roads could not have foreseen." They would not delay implementation of the law because of a shortage of telegraphers, "if it appears that higher wages would secure as many as are needed" for railroad service. The commissioners also stated bluntly that they would not postpone the March 4 deadline simply "because compliance will be inconvenient or costly" for railroads. At the February 27 hearing, railroad representatives testified about their inability to meet the new provisions of the law. A Southern Railroad representative claimed that the company would need to hire 220 additional telegraphers to meet the terms of the new law. The Pennsylvania Railroad asserted that it would require more than seven hundred additional operators to avoid any disruption to its dispatching operations. Other railroad managers named similar figures during the course of the hearing.[37]

The following week, ICC commissioners rejected all petitions and argued that railroads had failed to establish "good cause" for postponement of the nine-hour provision as required by HOSA. They warned firms that they must be in compliance with the new law when it went in effect two days later or face fines and penalties. Railroad officials and the railroad trade press railed against the commissioners' decision and argued that their strict construction of the language of HOSA would have dire consequences for the rail industry by "[adding] tremendously to the cost of operation[s]" and by forcing firms to close telegraph offices and make other service reductions. Railroad employees, on the other hand, cheered the decision and pledged their cooperation in enforcing the nine-hour provision of the new law.[38]

Railroad firms continued to challenge the ICC's administration of HOSA despite their failure to delay implementation of the law itself. On March 3, 1908, a day before the law went into force, ICC commissioners served 1,237 railroads with an order requiring them to periodically report all instances of employees being on duty longer than the limits established by HOSA. A number of major lines, including the Erie, the Baltimore & Ohio (B&O), and the New Haven & Hartford Railroads, filed suit against the ICC in federal court challenging the constitutionality of the ICC's reporting requirements on the grounds that the requirements forced firms to provide evidence that could be used against them if they admitted to violating HOSA. The lawsuits gradually made their way through the federal courts until the US Supreme Court ruled in *Baltimore & Ohio Railroad Company v. Interstate Commerce Commission* that the ICC had the constitutional authority to compel railroads to report on employee violations of HOSA. Following the 1911 decision, 153 firms that had steadfastly refused to file reports with the ICC began to comply with the original 1908 order.[39]

At the same time that railroad officials were attempting to delay implementation of HOSA and challenging the ICC's reporting provision, they were also pursuing short-term organizational strategies to mitigate the impact of the law on their rail operations. Across the nation, firms closed hundreds of remote telegraph facilities along lightly trafficked lines to free telegraphers for service at busier facilities. One down side of this practice was that dispatchers could no longer monitor train movements on these remote lines, which increased the risk of accidents and collisions.[40]

A number of major trunk lines that faced the biggest impact because of the nine- and thirteen-hour telegraph provision of HOSA, including the Pennsylvania, New York Central, Union Pacific, Southern Pacific, and the Santa Fe Railroads, established telegraph schools during the summer and fall of 1907 to train additional operators to man wayside stations and dispatching facilities. The schools offered six- to eight-month courses of study that familiarized students with basic Morse telegraphy and taught them other skills necessary for railroad service. The editor of the *Railroad Gazette* praised this new training program as a businesslike method for obtaining additional telegraphers without having to depend on the ORT's apprenticeship method. At the same time, the editor cautioned railroad officials that they could not expect to retain energetic and able telegraphers if they did not increase railroad wages to a level comparable with wages in other fields such as the commercial telegraph sector. The new training practices also created tensions between railroad officials and ORT members who believed—with some justification—that managers were trying to undermine ORT locals by employing nonunion, telegraph-school graduates.[41]

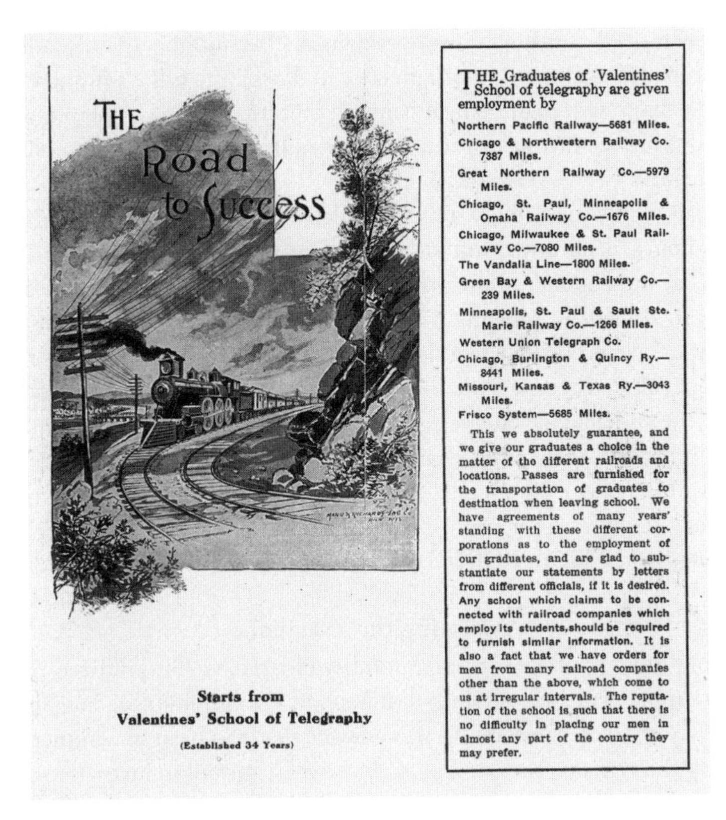

Numerous private telegraph schools were established in the second half of the nineteenth century to provide technical and practical training to students interested in pursuing careers as commercial and railroad telegraphers. Traditionally, a working knowledge of telegraphy had been acquired through apprenticeships with experienced telegraphers. Protective organizations such as the Order of Railroad Telegraphers attempted to maintain the apprenticeship system, but industry demand for telegraphers outstripped supply and led entrepreneurial telegraphers to establish training schools, such as Valentine's School of Telegraphy, which was founded in Janesville, Wisconsin, in 1870. *The Schwantes-Greever-Nolan Travel and Transportation Ephemera Collection, St. Louis, MO*

Finally, numerous firms attempted to modify HOSA's nine-hour time provisions for telegraphers by using so-called broken tricks, or split shifts, under which an employee would work a total of nine hours split up over a longer period of time. For instance, the employee might be required to be in the telegraph office from seven o'clock in the morning until seven o'clock in the evening but work only nine hours. Employees disliked broken tricks because their off-duty periods were often split into small time increments that could not be used effectively for rest or other nonwork activities. On a number of railroads, telegraphers rebelled against the work practice and used the ORT's influence to secure agreements with management banning split tricks unless employees received continuous breaks during longer working periods (e.g., a three-hour continuous break during a twelve-hour shift).[42]

The rail industry's inability to block or delay implementation of the nine- and thirteen-hour federal work limit for telegraphers and the ICC's unwavering commitment to uphold and enforce the terms of HOSA placed the American rail industry in an increasingly precarious position. The ORT's grassroots strength and political clout at the federal level prevented the railroad industry from attempting to amend or repeal the law. Short-term organizational responses by railroads to the anticipated acute telegrapher shortage once the law went into force offered limited relief for firms that depended on intensive train dispatching to keep their lines operating smoothly. Stopgap measures, however, did little to resolve the underlying labor conflicts and the telegrapher shortage that predated the passage of the Hours of Service Act.

Adopting the Telephone

Beginning in 1907, officials from a number of large, well-capitalized railroads began to implement a costly, long-term technical solution to the pressing labor crisis. They drew on a communication device that had been in commercial use for nearly thirty years, the telephone. The American rail industry's struggle to adopt the telephone as a train dispatching and operational management tool in the first decade of the twentieth century bore striking similarities to its struggles to implement telegraphic train dispatching more than a half-century earlier.[43] In both cases, managers initially dismissed or distrusted the new device and felt that it did not offer an improvement over older methods of operational management. They were also hesitant to modify established managerial procedures to accommodate the new communication tool. Ultimately, external political and economic pressures forced officials to act decisively. They began to investigate the experiences of early adopters within the rail industry and learned to overcome their inherent distrust in the time-tested communication device.

The telephone was not a completely unknown device in railroad circles when CB&Q officials in 1907 performed the first intensive telephone dispatching tests along the railroad's busy main line west of Chicago. However, only a handful of rail industry officials had acquired any practical firsthand experience using the telephone for dispatching purposes during the previous three decades. For the most part, railroad officials managing busy trunk lines were extremely hesitant to replace their expensive and sizable telegraph infrastructure with new and untested railroad telephone equipment.[44] In 1908, an editor for the *Railway Age* summed up the conservative attitude that had dominated the rail community over the past thirty years by noting that many officials felt that the telephone "might be on the whole as good as the telegraph, [but] it would not be any better; and the telegraph was established and was [therefore] satisfactory."[45] Only as the telegraph became less suitable for dispatching purposes owing to the work limits imposed by HOSA, and the concomitant labor shortage, did many officials begin to change their minds about the telephone.

A few innovative railroad officials had adopted the telephone for dispatching purposes in the early 1880s, but these experiments were limited to a handful of small and relatively short rail lines that had not previously used the telegraph for train dispatching and thus were not wedded to it. In the first such test of the device, Charles A. Hammond, superintendent of the Boston, Revere Beach & Lynn Railroad, installed telephones (almost certainly provided by Bell Telephone) on his busy commuter line in 1880, four years after Alexander Graham Bell and Elisha Gray introduced their respective telephone models to the public. Hammond later recalled that early railroad telephones had a host of problems. Inductive interference due to the proximity of poorly insulated telephone lines led to cross talk between the lines, as well as a "multitude of strange noises more or less confusing both to the sender and hearer of the message." Despite these problems, Hammond had found that telephones, "when surrounded by proper safeguards," were just as feasible as Morse telegraphy for train dispatching. He concluded that the telephone offered superintendents "a most helpful means of obtaining quickly a bird's-eye view, as it were, of the whole [operational] situation" on their divisions. Ironically, British telegraph pioneer William F. Cooke and Erie Railroad superintendent Daniel C. McCallum had both made exactly the same claim for telegraphy in nearly the same language half a century earlier.[46]

In 1883, divisional engineer, and later superintendent, Samuel Whinery of the New Orleans & Northeastern Railroad had a telephone line with Bell Telephone Company instruments installed along the right-of-way of the railroad's Northern Division to communicate with officials involved in the construction of the route. When train service began along the one-hundred-mile rail line, the telephone was used to transmit orders for the four daily trains that ran on the route. Whinery had agents transmit telephone orders in the same manner as telegraphed train orders, with the exception that all numbers and important words were spelled out as well as spoken. Telephone train dispatching continued for about a year until telegraph equipment was added to the line and senior officials ended the telephone dispatching experiment.[47]

A number of other short-line railroads adopted telephone train dispatching in the early 1890s. These included the Chicago, Kalamazoo & Saginaw in 1890 and the Western New York & Pennsylvania Railway in 1896. Unlike busy eastern and midwestern trunk lines, these companies averaged between fifty and sixty miles in length and did not have a high volume of daily freight and passenger traffic. With few stations to link by telephone and only a handful of trains operating each day, short-line managers did not have to invest large sums of money in telephone equipment and staff training to implement telephone train dispatching on their lines.[48]

Larger railroads with long-standing financial and organizational investments in their telegraph infrastructures had much less interest in adopting telephone dispatching in the late nineteenth century. Pennsylvania Railroad officials considered using the telephone for dispatching trains in the early 1890s but decided against the practice. They were particularly concerned that station operators would fail to record the train orders that they received orally from dispatchers. In the event of a collision, officials would not be able to determine which party was at fault through written evidence. Oral dispatching by telephone ran contrary to the rigid managerial rules that officials had developed over the span of forty years to accommodate the telegraph's use as a dispatching tool.[49]

Not all officials with busy main-line railroads distrusted the telephone. Charles L. Selden, telegraph superintendent for the Baltimore & Ohio Railroad, argued that the telephone and the telegraph were more alike than most officials believed. In a paper read before an 1894 meeting of the Association of Railroad Telegraph Superintendents, Selden argued that contemporary railroad telegraph instruments left no permanent record of the messages that were transmitted through them, since telegraphers transcribed all messages by ear, but railroad officials had devised safety protocols to ensure that telegraphers consistently transmitted and received messages accurately. The same rules and regulations could be applied to dispatching trains telephonically. A few years later, another expert on electrical appliances and railroad safety offered a similar argument for the superiority of the telephone over the telegraph in railroad service. Like Selden, George Blodgett argued that the telephone, when equipped with proper safeguards, would far surpass the telegraph as a tool for railroad management, particularly in moments of crisis when time was of the essence. Unlike the telegraph, which required messages to be coded, transmitted, decoded, and formatted for delivery, the telephone would allow officials to issue orders directly to station agents and train crews when necessary and instantly clarify any confusion. Blodgett saw this as a vast improvement over contemporary communication methods.[50]

By the late 1890s, a small number of progressive railroad officials had begun to rigorously evaluate the telephone's potential as an operational management tool. In 1897, the Telegraph Committee of the PRR's Association of the Transportation Officers (ATO) addressed the question of telephone dispatching at the request of senior railroad officials. They presented a lengthy report summarizing the views of managers throughout the nation. The report confirmed that officials across the country were skeptical about using telephones to issue train orders along busy main lines. Senior Baltimore & Ohio and Illinois Central managers categorically rejected the use of telephones for transmitting train orders from dispatchers to station agents. Likewise, an official from the Southern Railroad noted: "We do not use telephones in any way for the movement of trains, and I would be very loath to recommend them. While the telephone is very useful [for office business], I am afraid it is very unreliable to handle train orders over." Managers from the San Antonio & Aransas Pass Railroad concluded, "The use of the telephone is not permitted on this [rail]road . . . as we do not consider it safe to do so, for the reason that in case of accidents the responsibility cannot be located."[51] These railroad officials assured PRR managers that the telegraph was the only safe and reliable dispatching tool for railroad operations.

The Telegraph Committee did find that some officials used telephones for moving trains in busy rail yards, or in emergency situations when telegraph lines failed because of bad weather or technical problems. The Pennsylvania Railroad's Lines West of Pittsburgh Division reported to the ATO that all rail yards between Pittsburgh and Chicago used telephones for issuing train orders to yard crews. Both the CB&Q and Norfolk & Western Railroads employed telephones in large yards for controlling the movements of switching locomotives. An official with the Chicago & North Western Railway remarked that the company occasionally used telephones for dispatching trains when severe midwestern weather knocked out their telegraph lines.[52]

After reviewing the opinions of officials from over thirty railroads, members of the PRR Telegraph Committee concluded that "objections [to the telephone] arise from a natural prejudice against such an innovation, and because no practical test seems to have been made for long distances." Committee members advocated a battery of experiments to study telephone dispatching in action and proposed specific safety rules, such as requiring operators to write out telephoned orders on standard PRR "train-order" forms and then repeat the orders back to dispatchers for verification. The committee asserted that these minor adaptations of the Pennsylvania's long-established rules for recording train orders transmitted via the telegraph would eliminate any concerns that telephone dispatching would not leave a paper trail.[53]

Senior PRR officials considered the Telegraph Committee's recommendations but did not implement any of them, based on the advice of the PRR's Legal Department. At the time, a number of lawsuits related to the microphone patent granted to Emile Berliner in 1877, and subsequently acquired by American Bell, were making their way through the federal courts. Given the uncertain legal climate surrounding the ownership of a critical telephone patent, PRR officials chose to await the outcome of the court case before making any decisions about purchasing telephone equipment to conduct dispatching experiments.

This action echoed the company's decision in the early 1850s to delay the acquisition of a license to operate a telegraph network due to the acrimonious patent dispute between Samuel F. B. Morse and Royal House. A US circuit court ultimately voided aspects of the Berliner patent in 1901, but by that time, the topic of telephone train dispatching had ceased to be a priority for senior PRR officials. Two more years would pass before PRR officials resumed their investigation of the telephone's potential as a dispatching tool.[54]

The pages of industry trade journals such as the *Railroad Gazette* and the *Railway Age* document railroad officials' growing interest in the telephone between 1901 and 1907, but they also highlight the myriad of excuses offered by officials as to why the telephone was not gaining popularity as a dispatching tool on railroad main lines. Some authors cited the cost of telephone equipment and argued that the device was not yet reliable enough for railroad service. Others blamed railroad telegraphers for resisting efforts by officials to conduct experiments with telephone dispatching. Some felt that automatic block signaling would ultimately replace telegraphed train orders on main-line railroads and officials would be foolish to invest capital building telephone networks in the interim. A few argued that railroad employees were not disciplined enough to use the telephone for dispatching. Many observers, though, simply asserted that the telegraph remained the best tool for managing train movements because railroad officials had perfected the art and science of railroad telegraphy and train dispatching over the past half century.[55]

While the telegraph continued to be the dispatching device of choice for railroad firms operating trains along busy main lines in the years leading up to 1907, the telephone gained popularity as an adjunct to existing telegraphic train dispatching protocols on lines operated by the B&O and the Delaware, Lackawanna & Western Railroads. The B&O installed telephone booths along unmanned sidings between stations staffed with telegraph operators. Train crews could use the telephone to contact the telegraph operator at the nearest station and receive updated train orders. This meant that fewer telegraph operators were required along the B&O's many secondary routes and proved so successful that B&O officials authorized its expansion to all main-line sidings in 1906.[56]

Single-track branch lines, in particular, became a popular location for telephone installations. The PRR, CB&Q, B&O, and New York Central (NYC) all experimented with telephone dispatching on single-track branches. Lightly trafficked branch lines often served remote communities, and since the 1890s railroad firms had struggled to find sufficient numbers of competent telegraphers to staff out-of-the-way stations. The problem became acute owing to the growing shortage of skilled telegraphers during the first few years of the twentieth century. Telephone installations helped mitigate the shortage, since local residents with no training as telegraphers could be employed in these remote communities as telephone dispatchers and station agents.[57]

Despite the growing use of the telephone along branch lines, and as a supplement to telegraphic dispatching on main lines, many officials remained hesitant about installing new telephone equipment and wiring across their entire rail network. As an expedient, some firms installed so-called composite or telegraphone equipment on existing telegraph lines. The composite equipment permitted voice communication on telegraph lines. The sound quality of these composite lines was typically quite poor, since they utilized iron telegraph wires and employed a ground return instead of a metallic circuit.[58] Officials also found that composite networks were not reliable for long-distance connections along circuits more than two hundred miles in length. Nonetheless, they offered a bridge between the telegraph and the telephone and allowed managers to experiment with voice communication without having to spend large sums of money for specialized telephone equipment and high-quality copper wiring.[59]

The passage of the Hours of Service Act in March 1907 forced railroad telegraph superintendents and other senior railroad officials to evaluate in earnest whether telephone equipment had matured to the point where it could entirely replace the telegraph for dispatching trains on heavily trafficked, multitrack main-line corridors. Reducing the number of telegraph offices on these busy routes as a temporary expedient was simply not an option, and officials feared that the stringent terms of HOSA would leave them with an insufficient number of telegraphers to keep busy dispatching centers and stations staffed around the clock.

While senior railroad officials appealed to Congress and the ICC to modify or repeal HOSA in the fall of 1907 and spring of 1908, telegraph superintendents began quietly conducting experiments with telephone dispatching on a few busy main lines. New York Central and CB&Q superintendents took the lead. The NYC installed commercial telephone equipment along a forty-six-mile stretch of main-line track in September 1907. The only modification to the commercial, off-the-shelf telephone equipment involved adding an additional wire to the circuit that allowed dispatchers to selectively ring bells at specific stations to alert individual station operators to receive a train dispatch.[60]

The CB&Q followed suit in December with a major telephone test on its busiest stretch of track. CB&Q Telegraph Superintendent W. W. Ryder, acknowledged by his peers as the "father of telephone train dispatching," spearheaded the ambitious experiment. In the test, Ryder replaced all telegraph equipment with telephone receivers on a forty-six-mile stretch of the company's busy, multiple-track main line west of Aurora, Illinois.[61] In an article written six months after the test, Ryder noted that the results of the experiment had been "so satisfactory" that he converted more than three hundred miles of single- and double-track main line to telephone dispatching by the time HOSA took effect in March 1908. Ryder praised the convenience of communicating via the telephone: "The dispatcher is enabled to get far more detailed information of exactly what each train is doing. . . . He is thus brought just so much nearer the actual details of train movement." Ryder pointed out that railroad personnel with experience running trains made very effective telephone dispatchers: "A good, bright, young freight conductor . . . would make a better dispatcher . . . than the telegraph dispatcher, who, in spite of his occasional trips on a freight train is, after all, a theorist." Ryder discovered unexpected benefits to the telephone as well. Dispatchers and operators worked more efficiently under the new arrangement because superintendents could readily eavesdrop on their work. Telephones performed better than telegraphs in poor weather conditions. Ryder also noted that disabled railroad personnel could easily become telephone operators, since the position required little training beyond familiarity with railroad work. He concluded, "The field for substituting the telephone for the telegraph daily opens up before us almost faster than we can comprehend it, and the results we are obtaining from our experiments are a constant but very agreeable surprise."[62]

Ryder was not the only telegraph superintendent pleasantly surprised by the results obtained from telephone dispatching experiments on busy main lines. J. B. Fisher, the Pennsylvania Railroad's telegraph superintendent, praised the telephone's attributes in a letter to senior PRR officials in the spring of 1908. He noted that voice communication offered tremendous flexibility. The company would no longer have to employ highly skilled telegraph operators to transmit and receive train orders. They could cultivate anyone "with a knowledge of practical railroad work" into a competent telephone train dispatcher. This would eliminate labor shortages and give the railroad the upper hand in its dealings with the aggressive Order of Railroad Telegraphers.[63]

The results of the telephone experiments conducted by various railroads during the winter of 1907 and spring of 1908 were so successful that some industry officials began to predict both the demise of the telegraph as a train dispatching tool and the collapse of the ORT as an effective labor union. In August 1908, the editor of the *Signal Engineer*, the trade journal for railroad telephone and signaling matters, noted that fifteen railroads had installed telephone circuits along more than five thousand miles of track in the past six months and anticipated that the trend would cause railroad telegraphers to completely disappear from the industry within five years. He viewed this outcome as an ironic, yet fitting, result of the ORT's campaign to implement HOSA in 1907. At the same time, he failed to note an additional irony, namely, that railroad officials had fought vigorously to prevent passage of the law and owed the ORT and the federal government a debt of gratitude for spurring technical change within the industry through the new, stringent safety regulations.[64]

Telephone Triumphant

The technical transformation set in motion by the Hours of Service Act continued to gather momentum in the late 1900s and early 1910s. A January 1910 editorial in *Signal Engineer* revealed how rapidly managers on major trunk lines had incorporated telephony into their operations. The editor noted that PRR officials were in the process of equipping their busy main line between Altoona and Philadelphia with telephones. Other PRR divisions already had installed telephones for train dispatching. Train operations along nearly five hundred miles of main-line track throughout the central and eastern parts of the state were being managed through telephones. Based on this evidence, the editor concluded, "Now that [the telephone's] performances have satisfied the conservatives, its conquest is well-nigh complete."[65]

Telephone dispatching increased by leaps and bounds throughout 1909 and 1910. By June of 1910, barely two years after HOSA went into force, the editor of the *Signal Engineer* reported that 294 American railroads operated over twenty-six thousand miles of track (approximately an eighth of all track in the nation) through telephone dispatching. The CB&Q alone handled 2,437 miles of track by telephone, the Illinois Central 1,706 miles, and the Pennsylvania Railroad 1,100 miles. While many officials had initially viewed telephones, in the words of the *Signal Engineer* editor, as "the lesser of two evils" when compared to the prospect of doubling their workforce of telegraphers, they now recognized that telephones possessed "practically all the good qualities of the telegraph and many other qualities long desired but not previously obtainable." ICC officials also began to take note of the trend and launched an investigation in 1909 into the relative safety merits of the telephone versus the telegraph. While government officials could not offer a definitive assessment of the telephone's utility for railway dispatching in their December 1909 annual report, they concluded (much as railroad officials had) that the device itself appeared sound and that safety concerns could be resolved through proper training and managerial oversight.[66]

For railroad officials, one of the most important "long desired" qualities of the telephone was its potential to challenge the ORT's influence over the train dispatching field. Indeed, ORT membership peaked at approximately thirty-seven thousand members in 1908, shortly after enforcement of HOSA began. As more firms began to employ the telephone for train dispatching over the next few years, the union struggled to maintain its influence and relevance within the railroad industry. ORT leaders campaigned against efforts by railroads to hire nonunion employees as telephone operators and dispatchers at significantly lower wages than unionized telegraph operators and dispatchers. They also tried to force railroads to employ existing union members as telephone operators with

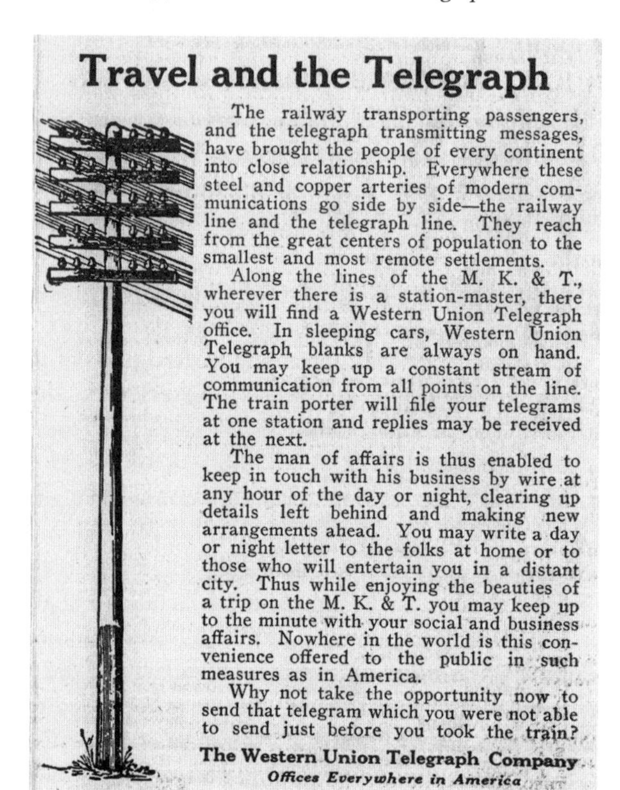

Travel and the Telegraph

The railway transporting passengers, and the telegraph transmitting messages, have brought the people of every continent into close relationship. Everywhere these steel and copper arteries of modern communications go side by side—the railway line and the telegraph line. They reach from the great centers of population to the smallest and most remote settlements.

Along the lines of the M. K. & T., wherever there is a station-master, there you will find a Western Union Telegraph office. In sleeping cars, Western Union Telegraph blanks are always on hand. You may keep up a constant stream of communication from all points on the line. The train porter will file your telegrams at one station and replies may be received at the next.

The man of affairs is thus enabled to keep in touch with his business by wire at any hour of the day or night, clearing up details left behind and making new arrangements ahead. You may write a day or night letter to the folks at home or to those who will entertain you in a distant city. Thus while enjoying the beauties of a trip on the M. K. & T. you may keep up to the minute with your social and business affairs. Nowhere in the world is this convenience offered to the public in such measures as in America.

Why not take the opportunity now to send that telegram which you were not able to send just before you took the train?

The Western Union Telegraph Company

Offices Everywhere in America

The telephone quickly replaced the telegraph as a tool for train dispatching following the 1907 Hours of Service Act, as railroads struggled to find sufficient numbers of telegraphers for their dispatching operations. Nevertheless, telegraph poles and wires continued to be a common sight along railroad rights-of-way throughout much of the twentieth century. Western Union employed these lines for commercial telegraph purposes and advertised the convenience and ubiquity of its telegraph services to railroad travelers, as this 1915 advertisement from a Missouri-Kansas-Texas Railroad timetable illustrates. *The Schwantes-Greever-Nolan Travel and Transportation Ephemera Collection, St. Louis, MO*

the same pay scales and seniority rules that telegraph operators enjoyed. Neither effort was entirely successful. ORT membership soon entered a period of decline that lasted until the outbreak of the First World War.[67]

Telephones and various manual and automatic electrical signaling devices largely replaced routine telegraphic train dispatching on busy, multitrack trunk lines by the beginning of the First World War. Though most American railroads continued to maintain telegraph networks for routine business communication and Western Union message traffic, few depended solely on telegraphy for operational management and train dispatching purposes. Telephone dispatching offered greater safety and efficiency, particularly as rising labor costs and the increasing scarcity of skilled telegraphers dissuaded companies from maintaining extensive manual telegraphic networks.

Despite the dramatic changes that took place in the railroad dispatching and signaling community during the half-decade that followed the implementation of the Hours of Service Act, industry predictions of the telegraph's—and the railroad telegraphers'—pending demise were greatly exaggerated. Well into the 1950s and 1960s, some railroad officials continued to perceive familiarity with Morse code as a basic prerequisite for railroad employment. The ORT survived the transition from telegraphy to telephony and, despite significant structural changes in the railroad industry such as the growing importance of centralized traffic control and the decline of commercial telegraph message traffic, continued to represent telegraphers and other affiliated railroad communication workers into the post–World War II era.

The factors that led to the passage of the Hours of Service Act in 1907, and rail industry officials' response to the legislation, highlight the important role that economic and political forces played in prompting technical and managerial change within the industry. Despite officials' concerns in the 1890s and early 1900s about the reliability of the "fallible guardians" manning telegraph keys in stations and dispatching centers across the nation, managers chose not to use economic and social incentives such as higher pay and better working conditions to secure a more dependable workforce. Furthermore, conservative officials were hesitant about supplementing or replacing telegraphic dispatching practices with alternative communication methods such as the telephone, owing to cost factors and concerns that new equipment might disrupt the standardized, interfirm management practices that officials had spent more than two decades developing. The small number of managers who had experimented with the telephone as a dispatching tool in the nineteenth century had done so in obscurity, and their practical demonstrations of the utility of the telephone for train management were largely overlooked by the rail industry until the early twentieth century. The industry's tremendous investments in telegraphy—both financial and organizational—during the second half of the nineteenth century precluded any widespread interest in replacing the telegraph, despite the potential of new communication devices to alleviate growing labor problems and safety concerns.

The political alliance between progressive reformers and labor organizers that led to the passage of the Hours of Service Act placed railroad officials in a situation that could not be resolved through short-term organizational and managerial reforms. The ORT's efforts to limit the entry of new telegraphers in the railroad workforce, at the same time that rail networks were continuing to expand during the first decade of the twentieth century, meant that officials could not simply hire more telegraphers, regardless of their skill level, to meet the stringent terms imposed by HOSA. In-house telegrapher schools offered a stopgap measure that did little to address the broader problem. The ORT's political strength prevented the industry from modifying or delaying implementation of the act. Consequently, rail officials were forced to pursue a drastic course of action and begin implementing telephone dispatching on busy main lines. Despite the costs and risks, the outcome of the telephone tests was so positive that rail officials embraced the new device as a means for combating both the telegrapher shortage and the strength of the ORT. Within two years, a significant percentage of rail mileage in the United States had been placed under telephone dispatching, and, by the mid-1920s, a majority of the railroad mileage would be dispatched via the telephone.[68]

Striking similarities exist between the rail industry's adoption of telephone dispatching and its adoption of telegraph dispatching in the 1850s and early 1860s. In both instances, rail officials struggled for years to determine how to adapt a device developed for commercial communication purposes to specialized railroad applications. In some instances, legal conflicts between the developers of rival telegraph patents had to be resolved before railroad officials were willing to invest financial resources in one device or another. Ultimately, the industry as a whole took a "wait and see" approach to both of these communication devices and committed financially and managerially only when faced with external economic and political pressures, and only after closely scrutinizing the experiences of early adopters. In the case of the telephone, HOSA provided the necessary impetus to force officials to confront the "fallible guardians" in their telegraph offices and seek a new communication tool for managing rail operations throughout the nation.[69]

Conclusion

By the end of the nineteenth century, the American railroad and telegraph industries had matured into national networks. Railroad companies grew from small lines serving local markets into major corporations that routinely operated thousands of trains across tens of thousands of miles of track. Similarly, telegraph promoters expanded Morse's forty-mile experimental telegraph line along the B&O Railroad's Washington Branch into a dense communication network that spanned the continent. Despite the success of both industries, at no point during the century did they establish unwavering financial and managerial ties with each other. Unlike rail networks in the United Kingdom, France, and the German states, American railroads were not stressed by traffic volume to a great extent during their formative period prior to the Civil War. Therefore, the first generation of American railroad managers simply had no compelling reason to incorporate telegraphy into railroad operations in the 1840s, unlike officials in these other countries. American railroad managers had developed safe, efficient, and inexpensive operating practices that could accommodate the moderate levels of train traffic found on American rail lines. Managers gradually improved these rule- and time-based operating protocols in the 1840s by standardizing regional timekeeping practices and improving employee access to accurate clocks and watches, rather than implementing major dispatching reforms. These incremental organizational and technical solutions met the immediate needs of operating personnel and were far less expensive and complex than adopting electrical telegraphy, which required negotiations for operating licenses, held the potential to expose railroads to legal liabilities, challenged traditional managerial hierarchies, and introduced new and unforeseen operating risks.

American railroad managers grudgingly introduced telegraphic train dispatching on their rail lines as traffic conditions during and after the Civil War pushed traditional operating methods to the breaking point. However, unlike in European nations, where the state had nationalized private telegraph firms, and, in some cases, entire railway networks, during the second half of the nineteenth century, American railroad managers were forced to maintain a relationship with a private, profit-maximizing, market-dominant carrier—Western Union—and a host of smaller firms that sought to challenge WU's dominance of the nation's communication networks. Jay Gould's efforts to construct a national telegraph network rivaling Western Union in the late 1870s had given some railroad officials hope that new competition might restore some balance to their relationship with the telegraph industry. A few major firms made common cause with Gould after he created the American Union Telegraph Company following the passage of the Butler Amendment in 1879. However, once Gould's long-term strategy succeeded and he acquired control of WU in 1881, railroad officials quickly discovered that the new Western Union differed little from the old Western Union.[1] They continued to deal with a firm that prioritized commercial telegrams over railroad-related business and thus impeded efforts to promote more efficient management through intensive use of the telegraph for train control.

Far from being natural and inherent partners, as some contemporary observers and a later generation of historians asserted, the two industries were drawn together through external factors such as war, increasing safety regulation at the state and national levels, and financial necessity on the part of the young telegraph industry, rather than because railroad and telegraph officials perceived their industries as inherently linked or beneficial to each other. Telegraph promotors with the firms that eventually merged into Western Union took advantage of many railroad officials' growing interest in access to telegraphy during and after the Civil War and locked railroad firms into long-term right-of-way lease agreements. Such arrangements protected Western Union by creating major barriers to entry for other telegraph firms. They also prevented railroads from seeking out alternate telegraph providers and forced them to rely on poor quality telegraph service provided by Western Union. This, in turn, limited their ability to implement more intensive telegraphic train management practices over time. For many railroad officials, the telephone was appealing as a dispatching tool at the turn of the twentieth century precisely because it was independent of Western Union's (and the Order of Railroad Telegraphers') control and gave railroad officials more power to structure their own internal communication networks as they saw fit. While telegraph lines remained a ubiquitous sight along railroad rights-of-way throughout the first half of the twentieth century, these lines carried an increasingly minor share of train dispatching traffic compared to telephone and later radio networks.

The idealized, allegorical image of railroad and telegraph unity that artist John Gast depicted in *American Progress* in 1872 overlooked years of conflict between railroad officials and telegraph promoters and did not hint at the frustrations and public policy decisions that would lead many railroad managers to cast off telegraphy in favor of telephone dispatching in the early twentieth century. Gast's representation of the transformative power of the steam locomotive and the electrical telegraph reflected deterministic cultural notions of American industrial and social progress and conveniently ignored the complex and problematic financial and operational relationship between the two industries.

While Gast as an artist may be pardoned for eschewing this contingent and multifaceted relationship in favor of a Whiggish depiction of historical progress, modern scholars must recognize that railroad officials and telegraph entrepreneurs did not simply discover *ex novo* that "each industry had something of value to offer the other."[2] The relationships that developed between these industries—financial, technical, and organizational—were the result of deliberate choices by middle- and upper-level managers, often acting in response to economic, political, and social forces outside their control. Railroad officials pursued rational courses of action that they believed to be in the best financial and legal interests of their firms over a time period spanning more than half a century. The reification of "technology" as an analytical lens through which to interpret change over time within institutions or infrastructure and the historical emphasis on "managerial capitalism" as a driving force for corporate growth in the second half of the nineteenth century have contributed to denying agency to these individuals by ignoring the much more prosaic and mundane internal and external factors that shaped their decision making.[3] The railroad and telegraph industries certainly influenced the emergence of new forms of capitalism in the late nineteenth century and the complex transportation and communication infrastructures that underpinned them, but railroad and telegraph officials were often reacting to broader political-economic changes in the United States, not necessarily driving those changes themselves. Ultimately, far from being "Siamese twins of commerce," in the words of a former Western Union executive, the railroad and telegraph industries were uneasy bedfellows, as Pennsylvania Railroad president Alexander J. Cassatt and numerous other railroad executives discovered time and again during the nineteenth and early twentieth centuries.[4] Efforts by Western Union officials and generations of historians to depict the relationship as anything but fraught ignore the fundamental conflicts and compromises that shaped the bonds between these two institutions and their workforce of "fallible guardians."

NOTES

Preface

1. Somewhat ironically, Britain nationalized its commercial telegraph lines in 1868, whereas American lines remained under private ownership. See John, *Network Nation*, 117, 124–125.

2. Wolff, *Western Union*, 205–206.

3. By the 1880s, approximately 90 percent of Western Union's telegraph lines ran along railroad rights-of-way. Wolff, *Western Union*, 204.

4. "General Manager," "The Block System on Single Track Railroads," *Railroad Gazette*, October 27, 1893.

5. Estes, Federal Writers' Project Interview.

Introduction

1. See Howe, *What Hath God Wrought*, 1–7, for a thoughtful contextualization of the telegraph's place within the economic, political, social, and cultural order of the United States in the 1840s. Howe characterizes the early nineteenth century "as a time of a 'communications revolution,'" broadly defined, that gave shape to the era and which culminated in the practical application of electrical telegraphy in the United States in the mid-1840s. For more on the communications revolution as a historical concept, see John, "American Historians." Historian of technology Leo Max discusses at length the attitude of Emerson and his contemporaries toward the locomotive and other transformative mechanical agents of change in the nineteenth century in his classic work *The Machine in the Garden*. Also see Marx's critical 2010 essay regarding the intellectual origins of the modern concept of *technology*, a term that he argues is imbued with inherent ambiguity when used to refer to complex mechanical, organizational, and political-economic systems such as the railroad or telegraph. Marx argues that the term blurs "the boundary between the material (physical, or artificial) components of these large socio-technological systems and the other, bureaucratic and ideological components." Given its hazardous character, to use Marx's phrase, the term *technology* will not be employed in this study in reference to the railroad and telegraph. Instead, this work will deconstruct these complex systems in order to better understand the material, personal, and bureaucratic interactions that helped to shape their development over time. Marx, "Technology," 575.

2. Crofutt may also have included the image in his guides as a means of reassuring his affluent readers that transcontinental railroad travel was safe because the telegraph would guarantee that they would never be cut off from contact with the outside world.

3. Wiebe, *Search for Order*, xiv.

4. Field, "Magnetic Telegraph," 402, 406.

5. White, *Railroaded*, 2.

6. See Wolff, *Western Union*, 204–206.

Chapter 1 • Rights-of-Way

1. Silverman, *Lightning Man*, 178. Italics in original.

2. Silverman, *Lightning Man*, 174–175. See Beauchamp, *Invented by Law.*

3. Silverman, *Lightning Man*, 177; Kitchenside and Williams, *Two Centuries of Railway Signaling*, 23. Cooke and Wheatstone's indicator telegraph used multiple circuits to operate individual needles on a display. The needles would deflect a certain number of times to indicate individual letters. Their device required little training to operate but was much slower and more error prone than Morse's single-circuit telegraph utilizing his eponymous code.

4. Kitchenside and Williams, *Two Centuries of Railway Signaling*, 23.

5. Silverman, *Lightning Man*, 178–183.

6. Silverman, *Lightning Man*, 183. Currency conversion based on Williams, *Appleton's Railroad and Steamboat Companion*, 234. The cost in 2017 dollars would have been roughly $258,000–282,000. Inflation calculation based on Consumer Price Index data from Measuring Worth, http://www.measuringworth.com/uscompare (accessed July 6, 2018). See Electro-Mechanical Train Locator, Samuel Finley Breese Morse Papers. Digital image courtesy of the Library of Congress, American Memory Project, Words and Deeds in American History: Selected Documents Celebrating the Manuscript Division's First 100 Years, reproduction no. A67, http://memory.loc.gov/mss/mcc/039/0001.jpg (accessed August 1, 2018).

7. Letter from Samuel F. B. Morse to B. F. Mudgett, October 23, 1857, quoted in Silverman, *Lightning Man*, 258.

8. Silverman, *Lightning Man*, 160–161. See John, *Spreading the News*; John, *Network Nation*, for a broader discussion of Morse's goals for his telegraph network.

9. Silverman, *Lightning Man*, 160–161.

10. Another factor that may have influenced Morse's decision is that he would need to negotiate a right-of-way agreement only with a single landowner, the railroad, rather than the multitude of private landowners that controlled the property along the public highways between Washington and Baltimore. The sources, however, are not explicit on this point.

11. Silverman, *Lightning Man*, 222. Technical problems and poor-quality materials forced Morse to abandon the underground design and resort to wires suspended from insulators on poles.

12. Reid, *Telegraph in America*, 753.

13. Dilts, *Great Road*, 63–64.

14. For a detailed analysis of the relationship between the United States military and American railroad development, see Angevine, *Railroad and the State.*

15. Stephen H. Long and William Gibbs McNeill, *Narrative of the Proceedings of the Board of Engineers of the Baltimore and Ohio Railroad Company, from Its Organization to Its Dissolution, Together with an Exposition of Facts, Illustrative of the Conduct of Sundry Individuals* (Baltimore, MD: Bailey & Francis, 1830), 34, quoted in, "United States Army," 150.

16. O'Connell, "United States Army," 152–154.

17. O'Connell, "United States Army," 155.

18. O'Connell, "United States Army," 174–177. Gerstner, *Early American Railroads.*

19. See Taylor, *Fast Life on the Modern Highway*, 184.

20. Yates, *Control through Communication*, 4. Historical sociologist Frederick C. Gamst has argued that these operating rules represented "the *sine qua non* of railroading." Without them, railroads could not have operated trains safely. Gerstner, *Early American Railroads.*

21. Shaw, *History of Railroad Accidents*, 22.

22. For a general overview of railroad development in New England during the nineteenth century, see Kirkland, *Men, Cities, and Transportation.*

23. Gerstner, *Early American Railroads*, 822.

24. Gerstner, *Early American Railroads*, 337–338.

25. Such liability rules became commonplace in the railroad industry. For instance, the New York & Erie Railroad's 1851 *Employee Rulebook* included the following dictate: "[Employees] will be liable to immediate dismissal for disobedience of orders, negligence or incompetency." *New York & Erie Rail Road, Instructions for the Running of Trains, Etc.* (New York: Parker's Journal, 1851), 1. Courtesy of the Railway and Locomotive Historical Society Collection, California State Railroad Museum.

26. For more on early railroad informational and technical networks, see Usselman, *Regulating Railroad Innovation*, 98–99.

27. Harlow, *Steelways of New England*, 134.

28. O'Connell, "United States Army," 218–220. William H. Smith's *Regulations* "may well have been the first actual [printed] *book* of rules on American railroads." Gestner, *Early American Railroads*, 822. Prior to Smith's volume, employee rules had been printed on the back of timetables issued to employees. Bartky, "Running on Time," 21.

29. O'Connell, "United States Army," 221. An internal investigation revealed that the conductor on one of the trains had ignored operational regulations and caused the collision.

30. O'Connell, "United States Army," 221–222. Also see Salsbury, *The State, the Investor, and the Railroad*, 183–190, for an overview of the public reaction to this and other accidents on the Western Railroad. In January 1842, the Massachusetts General Court launched an inquiry to determine if the state legislature should enact new regulatory rules to prevent future railroad accidents. The investigative committee concluded that the directors of the Western Railroad were not responsible for the October 1841 collision and had done their due diligence in devising safe operating protocols for the line. The committee recommended against new regulatory legislation and state oversight of railroad safety. Salsbury, *The State, the Investor, and the Railroad*, 189–190.

31. "Report on Avoiding Collisions and Governing the Employees," November 30, 1841, Western Railroad, 2–3, quoted in O'Connell, "United States Army," 222.

32. "Report on Avoiding Collisions and Governing the Employees," November 30, 1841, Western Railroad, quoted in Yates, *Control through Communication*, 5.

33. Yates, *Control through Communication*, 6.

34. Kieve, *Electric Telegraph*, 19.

35. Mott, *Between the Ocean and the Lakes*, 416.

36. Kieve, *Electric Telegraph*, 19–20.

37. Kieve, *Electric Telegraph*, 21.

38. Kieve, *Electric Telegraph*, 22–23.

39. Papers of William F. Cooke at the Institution of Engineering and Technology, London, UK (hereafter IET), vol. 1, September 8, 1837, quoted in Kieve, *Electric Telegraph*, 26.

40. Kieve, *Electric Telegraph*, 27. Approximately £23,370–29,210 per mile in 2017 British pounds. Inflation calculation based on Retail Price Index data from Measuring Worth, http://www.measuringworth.com/ukcompare (accessed July 6, 2018).

41. The British inventors later filed a patent in the United States on June 10, 1840, ten days before Morse filed his telegraph patent. The Cooke and Wheatstone patent was later

misdated, however, to 1842, possibly because of US patent commissioner Henry L. Ellsworth's intense desire to see an American receive credit for discovering the electrical telegraph. For a detailed discussion of Ellsworth's efforts to promote Morse and American inventors in general, see John, *Network Nation*, 47–48 and 434, no. 78.

42. IET, vol. 2, December 22, 1866, quoted in Kieve, *Electric Telegraph*, 29.

43. Kieve, *Electric Telegraph*, 30.

44. Cooke and Wheatstone's railroad telegraph lines would eventually see limited use by the British government and the press in the early 1840s, but they did not provide commercial telegraph service to the public. Cooke began work on the first commercial telegraph lines in the mid-1840s. Kieve, *Electric Telegraph*, 36–40. For another contemporary perspective on the relationship between the railroad and telegraph industries in Great Britain in the 1840s, see Murray, *Stokers and Pokers*.

45. Silverman, *Lightning Man*, 177.

46. Kieve, *Electric Telegraph*, 30–31.

47. Cooke, *Telegraphic Railways*, 1.

48. Cooke, *Telegraphic Railways*, 1.

49. Cooke, *Telegraphic Railways*, 2.

50. Cooke, *Telegraphic Railways*, 6.

51. Cooke, *Telegraphic Railways*, 10.

52. Cooke, *Telegraphic Railways*, 11.

53. For an authoritative political-economic history of the telegraph in the nineteenth century, see John, *Network Nation* and Hochfelder, *Telegraph in America*. For more on the transformation of the telegraph from a scientific and technical curiosity into an industrial tool, see Israel, *From Machine Shop to Industrial Laboratory*.

54. Reid, *Telegraph in America*, 753.

55. Silverman, *Lightning Man*, 224, 226.

56. Silverman, *Lightning Man*, 228. Smith's corrupt practices were legendary. One political opponent labeled him "'F.O.J. Smith, L.S.C.'—Liar, Scoundrel, and Coward."

57. The 44-mile-long line utilized approximately 1,160 poles, or roughly 6.6 miles worth of lumber. Such quantities of wood were easily obtained in the heavily forested American East and Midwest, but securing local supplies of timber in the much more arid West would prove challenging and would often require transporting supplies over great distances.

58. Silverman, *Lightning Man*, 236; Dilts, *Great Road*, 295–296.

59. "A letter from Professor Morse," Dec. 23, 1844, H.R. Document 24 (28-2), *U.S. Congressional Serial Set*, vol. 464 (Washington, DC, Government Printing Office, n.d.), 2.

60. Thompson, *Wiring a Continent*, 205.

61. John, *Network Nation*, 30–32, 38–40, 58–60. This arrangement remained in place until 1847, when the postmaster general began leasing the Baltimore-Washington line to the Magnetic Telegraph Company. Thompson, *Wiring a Continent*, 34.

62. Silverman, *Lightning Man*, 249–250.

63. Silverman, *Lightning Man*, 260–262.

64. Ely, *Railroads and American Law*, 13–16. Ultra vires literally means "beyond the powers" in Latin. *The Free Dictionary by Farlex*, http://legal-dictionary.thefreedictionary.com/ultra+vires (accessed July 6, 2017).

65. John, *Network Nation*, 68–75.

66. Silverman, *Lightning Man*, 262–265.

67. For financial reasons, Kendall attempted to build this section of the line before he began work on the Philadelphia-Baltimore link. He had contracted the construction of the line south of Philadelphia to Henry O'Rielly but had not lined up financial backers for the project. Reid, *Telegraph in America*, 114, 122.

68. Reid, *Telegraph in America*, 116–117.

69. Amos Kendall to Jane Kendall (née Kyle), June 6, 1845, quoted in "Railroads and the Telegraph," 7. Source originally from Stickney, *Autobiography of Amos Kendall*, 529.

70. Amos Kendall to Jane Kendall (née Kyle), June 6, 1845, quoted in "Railroads and the Telegraph," 7. Source originally from Stickney, *Autobiography of Amos Kendall*, 529.

71. John, *Network Nation*, 69–72; Reid, *Telegraph in America*, 116–117. Reid noted that a few years later, after patient negotiations, he successfully obtained a right-of-way contract from the C&A. Reid, *Telegraph in America*, 117.

72. See Camden & Amboy Railroad and Transportation Company, Minute Book of the Executive Committee and Joint Board of the Camden & Amboy and the New Jersey Railroad and Transportation Company, vol. 2, June 6, 1845, 302; October 15, 1845, 308; November 24, 1845, 307. Accession 1807: Pennsylvania Railroad Company, record group 3: Lines East Corporate Records, New Jersey Railroad & Transportation Company, microfilm reel 70: Pennsylvania Railroad Company Minute Books, Hagley Museum and Library.

73. Reid, *Telegraph in America*, 117.

74. See New Jersey Railroad & Transportation Company, Minute Book of the Board of Directors, vol. 2, April 16–June 20, 1845, 43–48. Accession 1807: Pennsylvania Railroad Company, record group 3: Lines East Corporate Records, New Jersey Railroad & Transportation Company, microfilm reel 70: Pennsylvania Railroad Company Minute Books, Hagley Museum and Library.

75. Kendall to Morse, August 5, 1845, Samuel Finley Breese Morse Papers, quoted in Nonnenmacher, "Law, Emerging Technology, and Market Structure," 73.

76. Nonnenmacher, "Law, Emerging Technology, and Market Structure," 67. The legislation applied only to publicly owned roads and waterways. Railroads and private turnpikes remained exempt.

77. "Railroads and the Telegraph," 7.

78. Thompson, *Wiring a Continent*, 80.

79. Philadelphia, Wilmington & Baltimore Railroad Company, Minute Book of the Stockholders and Board of Directors (hereafter PW&B Minutes), vol. 1, December 9, 1845, 423; and January 7, 1846, 424. Accession 1807: Pennsylvania Railroad Company, record group 3: Lines East Corporations, Philadelphia, Wilmington & Baltimore Railroad Company, vol. 236, Hagley Museum and Library.

80. PW&B Minutes, vol. 1, December 9, 1845, 423; and January 7, 1846, 424. Accession 1807: Pennsylvania Railroad Company, record group 3: Lines East Corporations, Philadelphia, Wilmington & Baltimore Railroad Company, vol. 236, Hagley Museum and Library.

81. Stephen Paschall to E. C. Dale, March 20, 1846, Philadelphia, Wilmington & Baltimore Railroad Company Board Papers (hereafter PW&B Board), B-1347, folder 20. Accession 1807: Pennsylvania Railroad Company, record group 3: Lines East Corporations, Philadelphia, Wilmington & Baltimore Railroad Company, Hagley Museum and Library.

82. Stephen Paschall to E. C. Dale, March 24, 1846, PW&B Board, B-1347, folder 20. Accession 1807: Pennsylvania Railroad Company, record group 3: Lines East Corporations, Philadelphia, Wilmington & Baltimore Railroad Company, Hagley Museum and Library.

83. E. C. Dale to Stephen Paschall, June 12, 1846, PW&B Board, B-1347, folder 20. Accession 1807: Pennsylvania Railroad Company, record group 3: Lines East Corporations, Philadelphia, Wilmington & Baltimore Railroad Company, Hagley Museum and Library. The Telegraph Act of 1866 would later grant telegraph companies the right to build telegraph lines on "all railways," but it did not address whether this required railroad permission. In the subsequent Pensacola Telegraph Case of 1877, the Supreme Court ruled that railroad firms could not be forced to open their rights-of-way to telegraph companies, though this precedent became less significant for both industries, owing to federal legislation in 1879. While some states permitted telegraph companies to utilize eminent domain to access railroad rights-of-way, federal courts continued to affirm railroad companies' right, with a few exceptions, to control access to their rights-of-way. For more information on the Telegraph Act of 1866, see John, *Network Nation*, 116–122; Wolff, *Western Union*, 204–205, 248, 253. For a detailed summary of nineteenth-century state and federal laws and court decisions related to telegraph company access to railroad rights-of-way, see Cooke, *Treatise on Telegraph Law*, 29–66.

84. PW&B Minutes, vol. 1, September 8, 1846, 453, and December 14, 1847, 504; vol. 2, April 13, 1852, 142. Accession 1807: Pennsylvania Railroad Company, record group 3: Lines East Corporations, Philadelphia, Wilmington & Baltimore Railroad Company, vol. 236, Hagley Museum and Library.

85. Reid, *Telegraph in America*, 353–354.

86. Reid, *Telegraph in America*, 354, 357.

87. Reid, *Telegraph in America*, 358.

88. "Singular Railroad Accident," *The Farmer's Cabinet*, April 1, 1847.

89. Nonnenmacher, "Law, Emerging Technology, and Market Structure," 76.

90. "The Telegraph," *Barre Gazette*, April 2, 1847.

91. G. A. Nicolls to J. D. Stude, May 7, 1847. Accession 1520: Reading Corporation, series 1, box 103, folder 4, Hagley Museum and Library.

92. Reid, *Telegraph in America*, 358, Nonnenmacher, "Law, Emerging Technology, and Market Structure," 75–76; "Railroads and the Telegraph," 8.

93. Reid, *Telegraph in America*, 359. Italics in original.

94. The New England press eagerly reported on the problems experienced by Smith's New York–Boston line. These reports were commonly copied by newspapers in other regions, thus broadening the audience who learned of the problems caused by his unreliable telegraph line.

95. Jeptha H. Wade, "Sketch of the Life of J. H. Wade from 1811 to about 1867," Jeptha H. Wade Papers, series 1, container 1, folder 2, 15, Western Reserve Historical Society (hereafter "Sketch of the Life").

96. "Sketch of the Life."

97. Bartky, *Selling the True Time*, 22. Alexis McCrossen briefly discusses this standardization effort in her study of timekeeping in the United States: McCrossen, *Marking Modern Times*, 47–48. For a broader discussion of time and timekeeping, see Stephens, *Time and Navigation*.

98. Bartky, "Running on Time," 22.

99. Bartky, "Running on Time," 23.

100. Bartky, "Running on Time," 22.

101. Bartky, "Running on Time," 23.

102. Shaw, "Railroad Accidents," 37.

103. Bartky, *Selling the True Time*, 22. By the end of 1849, the New England Association of Railroad Superintendents represented nearly half of the seventy operating companies in New England.

104. Bartky, *Selling the True Time*, 22–23. Bartky notes that the choice of Bond by the association was perplexing, since another Boston watchmaker, Simon Willard Jr., was already providing time readings for many Boston railroad firms. He further argues that the selection of a railroad time zone two minutes earlier than Boston civil time made little sense, given the number of railroad firms operating out of the city. Michael O'Malley asserts that Bond was selected as the official timekeeper for the association because of his reputation as a man of science and his involvement in various scientific endeavors of the era such as the construction of the Harvard Observatory in 1839 and his participation in longitude measurements for the United States Coast Survey. See O'Malley, *Keeping Watch*, and Stephens, "Astronomy as Public Utility."

105. Bartky, *Selling the True Time*, 235n8.

106. Reid, *Telegraph in America*, 357.

107. Reid, *Telegraph in America*, 378.

108. Bond eventually arranged for a telegraph connection with the Harvard Observatory in 1850. He contracted with the new Vermont and Boston Telegraph Company to connect his store with the observatory and transmit daily time signals. Bartky, *Selling the True Time*, 59. A year later, Boston railroad officials began to obtain time readings directly from the observatory through an automated system of Bond's invention. O'Malley, *Keeping Watch*, 66–67. Also see Ian R. Bartky's and Carlene Stephens's comments about their disagreements regarding telegraphy and railroad timekeeping in New England in the late 1840s and early 1850s. Bartky argues that Bond's telegraphed time signals were not revolutionary, since railroads had developed timekeeping and operating practices in the pretelegraph era that were precise enough for safe operations. Stephens asserts that Bartky misunderstands her argument; rather than telegraphed time signals being significant, she argues that Bond's automation of the time transmission process in 1851 by wiring a clock into the telegraph circuit itself was the revolutionary act. Bartky and Stephens, "Comment and Response."

Chapter 2 • Dangerous Expedient

Epigraph. Andrew Carnegie, *The Autobiography of Andrew Carnegie* (New York: Houghton Mifflin Company, 1920), 70.

1. Shaw, *A History of Railroad Accidents*, 120.
2. Reid, *Telegraph in America*, 160.
3. *Autobiography of Andrew Carnegie*, 59.
4. Historian Richard White argues that "Tom" Scott embodied both the best and worst characteristics of nineteenth-century railroad officials. On one hand, he was a thoroughly competent railroad manager with in-depth knowledge of rail operations. On the other hand, he was a skilled political manipulator who thought little of bribing state and federal politicians to secure legislation favorable to the Pennsylvania Railroad and later the Texas & Pacific Railway. See White, *Railroaded*.
5. The New York & Erie Railroad completed the first trans-Appalachian line across the Southern Tier of New York in April of 1851. Hungerford, *Men of Erie*, 111. Also see Churella, *Building an Empire*.

6. Wall, *Andrew Carnegie*, 115; *Autobiography of Andrew Carnegie*, 63.

7. Andrew Carnegie to Uncle George Lauder, March 14, 1853, Andrew Carnegie Papers, vol. 70; quoted in Wall, *Andrew Carnegie*, 117.

8. "Report of the Superintendent of Transportation," January 1, 1852, from *Fifth Annual Report of the Pennsylvania Railroad Company to the Stockholders*, 63–64. Accession 1807: Pennsylvania Railroad Corporation, record group 1, series 9, box 1421, Hagley Museum and Library.

9. *Autobiography of Andrew Carnegie*, 70.

10. *Autobiography of Andrew Carnegie*, 70–71.

11. Business historian Steven Usselman borrows the term *insider innovation* from Naomi Lamoreaux's writings. See Lamoreaux, *Insider Lending: Banks, Personal Connections, and Economic Development in Industrial New England* (Cambridge, UK: Cambridge University Press and NBER, 1996), quoted in Steven W. Usselman, *Regulating Railroad Innovation*, 65.

12. A notable variation on Morse code emerged outside the United States in the late 1840s. A German, Friedrich Gerke, derived a simpler version of the code by eliminating some of the long dashes and closely spaced dots used for certain characters, as well as changing the sequences of dots and dashes assigned to various letters and numbers. Though slower than traditional Morse code, the German code was simpler and thus better suited for use over submarine cables. American telegraphers resisted adopting Continental Morse, and the two variations (later referred to as International Morse and American or Railroad Morse) remained in use throughout the nineteenth and the first half of the twentieth centuries until American Morse code began to gradually fade away with the decline in manual telegraphy in the United States. See Coe, *Telegraph*.

13. Silverman, *Lightning Man*, 323.

14. Silverman, *Lightning Man*, 284–285; Thompson, *Wiring a Continent*, 54–55.

15. See Hungerford, *Men of Erie*, chaps. 2–10. The Erie was the first of the four major trunk lines to complete a route from the Atlantic tidewater to the Great Lakes and the Ohio River Valley. The Pennsylvania Railroad completed the second route in late 1852. The Baltimore & Ohio Railroad finished the third line early in 1853. Finally, investors created the New York Central in mid-1853 by merging twelve separate railroad companies that ran from Schenectady to Buffalo, New York.

16. *American Railway Times*, May 2, 1850.

17. George, *Forty Years on the Rail*, 28, 60.

18. Bartky, "Running on Time," 27. Also see Eastern Railroad Association, *Records of the New England Association of Railway Superintendents*. See chap. 1 for more information on the New England Association of Railway Superintendents.

19. *American Railway Times*, May 2, 1850.

20. Reid, *Telegraph in America*, 293.

21. Nonnenmacher, "Law, Emerging Technology, and Market Structure, 37.

22. Nonnenmacher, "Law, Emerging Technology, and Market Structure," 42.

23. Nonnenmacher, "Law, Emerging Technology, and Market Structure," 44; Reid, *Telegraph in America*, 289. Smith disliked Kendall and believed that he was trying to dominate the fledgling telegraph industry. The previous year, Smith had derailed Kendall's efforts to build a New York–Boston extension for the Magnetic Telegraph Company's line. Reid, *Telegraph in America*, 353–354. See chap. 1 for more information on Smith's New York–Boston line.

24. Mott, *Between the Ocean and the Lakes*, 415.

25. Nonnenmacher, "Law, Emerging Technology, and Market Structure," 45.

26. Reid, *Telegraph in America*, 192.

27. Mott, *Between the Ocean and the Lakes*, 430. Also see Bartky, "Running on Time," 27.

28. Mott, *Between the Ocean and the Lakes*, 415. The Erie's board of directors believed that the cost of paying telegraph operators would exceed any financial benefits to the railroad. Minot countered by arguing that station agents and clerks could be trained to operate the telegraph at no additional cost. Thompson, *Wiring a Continent*, 206–207.

29. Reid, *Telegraph in America*, 293.

30. Reid, *Telegraph in America*, 294.

31. Mott, *Between the Ocean and the Lakes*, 415.

32. Mott, *Between the Ocean and the Lakes*, 417. Minot later appointed Conklin as the first salaried telegraph operator on the Erie Railroad.

33. *New York and Erie Railroad: Instructions for the Running of Trains, Etc. to Go into Effect on Monday, March 31, 1851* (New York: Parker's Journal, 1851), 5. Courtesy of the Railway & Locomotive Historical Society Collection, California State Railroad Museum.

34. Erie historian Edward Hungerford gives the date as September 22, 1851, but provides no primary source evidence to support his assertion. Hungerford, *Men of Erie*, 93.

35. Mott, *Between the Ocean and the Lakes*, 420.

36. Mott, *Between the Ocean and the Lakes*, 420. Minot's experiment on the Erie was the most influential "first" use of the telegraph for train dispatching in the United States, though it may not have been the first time that a telegraph was used for issuing train orders. Charles Haskins, a conductor on the Michigan Central Railroad, later asserted that he had used the telegraph for a similar purpose during the winter of 1849–1850. Johnston, *Telegraphic Tales*, 181. In 1912, a monument to Charles Minot was dedicated at Harriman, New York, along the Erie main line. Funds for the monument were raised by current and former telegraphers across the nation, including Andrew Carnegie and Thomas Edison. Cassale, "Monument to Charles Minot." In 1951, for the centennial of the famous "first dispatch," the Morse Telegraph Club of America arranged for a new commemoration event at Harriman, New York, and Erie Railroad and Western Union officials participated in the celebration. "Centennial of Train Dispatching," 518.

37. Mott, *Between the Ocean and the Lakes*, 419.

38. Mott, *Between the Ocean and the Lakes*, 419.

39. Cornell personally purchased the telegraph franchise at auction and sold it to the Erie Railroad. He maintained a financial interest in the commercial portion of the line and renamed the company the New York and Western Union Telegraph Company. Nonnenmacher, "Law, Emerging Technology, and Market Structure," 46.

40. Mott, *Between the Ocean and the Lakes*, 415–416.

41. Shaw, *History of Railroad Accidents*, 29–30.

42. "Report of the Directors of the New York and Erie Railroad to the Stockholders," *American Railroad Journal*, vol. 27, January 14, 1854, 21.

43. "Report of the Directors of the New York and Erie Railroad to the Stockholders," *American Railroad Journal*, vol. 27, January 14, 1854, 21.

44. Ward, *J. Edgar Thomson*, 70. The name of the company was eventually shortened to the Pennsylvania Railroad.

45. Ward, *J. Edgar Thomson*, 71. Thomson began his career constructing railroads in the South. He worked his way up to the position of chief engineer on the Georgia Railroad during the early 1840s. See Ward, *J. Edgar Thomson*, chap. 2, for more information on Thomson's work for the Georgia Central Railroad.

46. Ward, *J. Edgar Thomson*, 71–76.

47. Ward, *J. Edgar Thomson*, 99.

48. For more details on Herman Haupt's professional training and early work for the PRR, see Ward, *That Man Haupt*, chaps. 1 and 2.

49. See Bartky, "Running on Time," 23–33 and 35, nn. 17, 22, 26.

50. Reid, *Telegraph in America*, 160, 165.

51. Pennsylvania Railroad, Minute Book of the Board of Managers (hereafter PRR Board Minutes), vol. 1, March 29, 1849, 156. Accession 1807: Pennsylvania Railroad Corporation, record group 1, series 6, Hagley Museum and Library.

52. PRR Board Minutes, vol. 1, November 21, 1849, 217.

53. "Report of the Superintendent of Transportation," January 1, 1852, from *Fifth Annual Report of the Pennsylvania Railroad Company to the Stockholders*, 63–64. Accession 1807: Pennsylvania Railroad Corporation, record group 1, series 9, box 1421, Hagley Museum and Library.

54. *Autobiography of Andrew Carnegie*, 63–64.

55. PRR Board Minutes, vol. 2, April 7, 1852, 153.

56. Ward, *J. Edgar Thomson*, 90; PRR Board Minutes, vol. 1, April 21, 1852, 157.

57. PRR Board Minutes, vol. 2, December 15, 1852, 280.

58. Pennsylvania Railroad, Special Committee of Board of Managers (hereafter PRR Special Committee), microfilm roll 27/35, April 6, 1853. Accession 1807: Pennsylvania Railroad Corporation, record group 1, series 8, Hagley Museum and Library.

59. PRR Special Committee, April 6, 1853.

60. PRR Special Committee, April 6, 1853.

61. PRR Special Committee, April 6, 1853.

62. PRR Board Minutes, vol. 2, April 6, 1853, 327; PRR Special Committee, September 12, 1853.

63. PRR Board Minutes, vol. 2, January 4, 1854, 422.

64. PRR Board Minutes, vol. 3, September 19, 1855, 102.

65. Reid, *Telegraph in America*, 174.

66. Alfred D. Chandler Jr. champions McCallum as a pioneer in the field of business management, particularly with respect to "perfecting the flow of internal information so essential to top and middle management to coordinate complex widespread activities and monitor and evaluate the performance of the large number of managers handling them." Chandler, *Visible Hand*, 101–105. Chandler conveniently overlooks McCallum's clashes with railroad personnel that eventually led to his resignation; this prevented him from fully implementing his complex data reporting protocol in which frontline personnel would keep upper-level managers constantly apprised of key operating metrics via the telegraph. As would again be demonstrated during McCallum's service as head of the U.S. Military Railroad during the Civil War, he excelled at devising innovative management strategies but struggled to implement them on a day-to-day basis. For more information on McCallum's Civil War railroad service, see chap. 3.

67. Mott, *Between the Ocean and the Lakes*, 431.

68. Mott, *Between the Ocean and the Lakes*, 434.

69. Mott, *Between the Ocean and the Lakes*, 431.

70. Mott, *Between the Ocean and the Lakes*, 431.

71. Mott, *Between the Ocean and the Lakes*, 431–432.

72. *New York and Erie Railroad Telegraph. Instructions for the Working of the Line* (New York: New York and Erie Printing Office, 1854, 7). Courtesy of the Railway and Locomotive Historical Society Collection, California State Railroad Museum (hereafter NY&E, *Instructions*).

73. NY&E, *Instructions*, 13–14, 18–19.

74. NY&E, *Instructions*, 14.

75. Mott, *Between the Ocean and the Lakes*, 421.

76. Mott, *Between the Ocean and the Lakes*, 421.

77. Mott, *Between the Ocean and the Lakes*, 421.

78. McCallum also devised and issued the first modern organizational diagram in his 1855 report. McCallum's diagram resembled a tree, rather than the pyramidal shape of later organizational charts. For a more detailed analysis of McCallum's organizational diagram, see Rosenthal, "Big Data."

79. Chandler, *Visible Hand*, 98.

80. Daniel C. McCallum, "Superintendents Report," March 25, 1856, in *Annual Report of the New York and Erie Railroad Company for 1855* (New York: New York and Erie Printing Office, 1856), 33–37, 39–41, 50–54, 57–59, quoted in Chandler, *Railroads*, 101.

81. Chandler, *Railroads*, 104.

82. Chandler, *Railroads*, 105.

83. Chandler, *Railroads*, 105.

84. Reid, *Telegraph in America*, 478.

85. See chap. 1.

86. Chandler, *Visible Hand*, 104. Alfred Chandler devotes significant attention to McCallum's 1855 report in *The Visible Hand*, as well as in *The Railroads*, his source book on the American railroad industry. He goes so far as to reproduce the 1855 report in the source book (Chandler, *Railroads*, 101–108). Chandler credits McCallum with discovering that "the telegraph was more than merely a means to make train movements safe. It was a device to assure more effective coordination and evaluation of the operating units under [McCallum's] command." Chandler, *Visible Hand*, 103. It should be kept in mind, however, that McCallum's views were far ahead of his time and did not reflect the general attitude of railroad officials toward telegraphy during either the antebellum or early postbellum years. Other railroad officials, particularly J. Edgar Thomson of the Pennsylvania Railroad, incorporated elements of McCallum's managerial principles into their organizational manuals, but no officials adopted them wholesale. Chandler, *Visible Hand*, 105. It is also important to acknowledge Richard White's warning that, "like so many railroad publications, the railroad organization chart was often a fiction. . . . [It] depended on managerial capacity—the ability to communicate orders and monitor their implementation—honesty, competence, and ultimately, intent. . . . It demanded subordination. . . . The charts often attributed capacity, honesty, and subordination that did not actually exist. . . . The railroad workforce was the product of 'patronage, favoritism, nepotism, and extortion.' There would be attempts to centralize, standardize, and control these organizations, but they had not succeeded by the 1890s." Walter Licht, *Working for the Railroad: The Organization of Work in*

the Nineteenth Century (Princeton, NJ: Princeton University Press, 1983), 58, quoted in White, *Railroaded*, 236.

87. Mott, *Between the Ocean and the Lakes*, 432–434. Following his resignation, McCallum worked as a structural engineer. He developed an inflexible arch truss for wooden railroad bridges that was warmly received by railroad engineers. McCallum and his family lived comfortably on licensing fees from the design until the beginning of the Civil War.

88. *Instructions for the Running of Trains, Etc. on the New York and Erie Railroad*, 1857, 15; Mott, *Between the Ocean and the Lakes*, 123–124. The New York and Erie eventually went into receivership in 1859, a victim of needless overexpansion and declining freight and passenger revenues.

89. Dilts, *Great Road*, 382.

90. Dilts, *Great Road*, 391; Reid, *Telegraph in America*, 755.

91. Reid, *Telegraph in America*, 313.

92. The board of directors of the NYCR may have been concerned about stockholders viewing the agreement between the executive officers of the two companies as a form of collusion or a violation of ultra vires.

93. Reid, *Telegraph in America*, 314; New York Central Railroad Company, *Annual Report*, 10.

94. Reid, *Telegraph in America*, 317.

95. Henry O'Rielly, "Railway Telegraph System: For Facilitating Business and Protecting Life and Property along Railroads, July 1852," quoted in Thompson, *Wiring a Continent*, 210.

96. Thompson, *Wiring a Continent*, 210.

97. For example, see G. A. Nicolls to John Tucker, March 6, 1855, Accession 1520: Philadelphia & Reading Railroad Company, series 1, box 104-5, Hagley Museum and Library (hereafter Reading Collection).

98. G. A. Nicolls to Franklin B. Gown, October 22, 1869, Reading Collection, series 1, box 105-3.

99. G. A. Nicolls to John Tucker, March 6, 1855, Reading Collection, series 1, box 104-5.

100. Bartky, *Selling the True Time*, 24–25.

101. See Bartky, "Running on Time," 29, for more details on the cause of the crash.

102. Bartky, *Selling the True Time*, 25–27.

103. Bartky, *Selling the True Time*, 27–28; Bartky, "Running on Time," 30.

104. Bartky, *Selling the True Time*, 25–26.

105. Reid, *Telegraph in America*, 235.

106. Reid, *Telegraph in America*, 237–238.

107. Illinois Central Railroad, Minutes of the Executive Committee (hereafter IC Executive Committee), vol. 25, October 24, 1853, 140. Illinois Central Railroad Papers, record group 3, series +3.1, Newberry Library.

108. Reid, *Telegraph in America*, 240; Corliss, *Main Line of Mid-America*, 65.

109. IC Executive Committee, vol. 26, December 20, 1854, 82–83.

110. *1855 Annual Report of the Illinois Central Railroad Company*, 7, quoted in Yates, *Control through Communication*, 106; Corliss, *Main Line of Mid-America*, 64. At the time of its completion, the nearly four-hundred-mile-long IC was the longest railroad in the world.

111. Yates, *Control through Communication*, 106; Reid, *Telegraph in America*, 244.

112. Overton, *Burlington Route*, 39–40; Chicago, Burlington & Quincy Railroad, "Contract between the Chicago and Mississippi Telegraph Company and the CB&Q and the Central Military Tract Railroad Companies" (hereafter CB&Q Contract), February 14, 1856. Chicago, Burlington & Quincy Railroad Papers, record group 33.1870.5.1, CB&Q Abstract of Telegraph Contracts, Newberry Library. Caton's contract with the CB&Q differed in only minor details from his contract with the Illinois Central.

113. Reid, *Telegraph in America*, 240.

114. Yates, *Control through Communication*, 106.

115. CB&Q Contract, April 15, 1863. Also see Usselman, *Regulating Railroad Innovation*, 126.

116. Reid, *Telegraph in America*, 245. Hughett later became president of the Chicago & North Western Railroad. Railroad and Warehouse Commission of Minnesota, *Twenty-First Annual Report*, n.p.

117. Thompson, *Wiring a Continent*, 214.

118. Read, *Telegraph in America*, 468–469.

119. Jeptha H. Wade, "Sketch of the Life of J. H. Wade from 1811 to about 1867," Jeptha H. Wade Papers, series 1, container 1, folder 2, 15, Western Reserve Historical Society (hereafter "Sketch of the Life").

120. "Sketch of the Life," 16.

121. Michigan Central Railroad Company, *Report of the Directors*, 5, 19. Wade noted in his memoir that Brooks raved about the usefulness of the Western Union line for train control purposes in his 1856 annual report, but he did not mention Brooks's critical comments about the importance of a railroad-controlled telegraph line. "Sketch of the Life," 16.

122. Reid, *Telegraph in America*, 470, 476, 479, 480–481. By the 1880s, approximately 90 percent of Western Union's telegraph lines ran along railroad rights-of-way, a direct result of Wade and Stager's expansion strategy in the 1850s. Wolff, *Western Union*, 204.

123. Reid, *Telegraph in America*, 243.

124. *Autobiography of Andrew Carnegie*, 69; O'Brien, *Telegraphing in Battle*, 42–43, 45.

125. O'Brien, *Telegraphing in Battle*, 45; *Instructions for the Running of Trains, Etc. on the New York and Erie Railroad*, 1857, 76. Operators at smaller stations were ordered to sign off promptly at 8:30 p.m. every evening and return to duty at 8 a.m. the following morning.

Chapter 3 • At War with Time and Space

1. Angevine, *Railroad and the State*, 132.

2. Annual Report of Daniel C. McCallum, May 26, 1866, *War of the Rebellion: Official Records of the Union and Confederate Armies* (Washington, DC: Government Printing Office, 1880–1901), series 3, vol. 5, no. 1, 1003 (hereafter *ORA*); Weber, *Northern Railroads*, 15; Plum, *Military Telegraph*, 380; Annual Report of Col. Anson Stager, September 15, 1865, *ORA*, series 3, vol. 5, no. 1, 361.

3. The PRR's telegraphic management practices utilized telegraphs located in stations along its single-track main line to report train locations and movements. Dispatchers could quickly resolve delays by issuing new schedules and operating instructions to train crews. This allowed for safe and efficient train movements when unforeseen circumstances prevented train crews from operating according to the railroad's strict timetable. See chap. 2 for more information.

4. See chap. 2 for additional details about McCallum's management practices.

5. General scholarship on Civil War transportation and communication history includes Plum, *Military Telegraph*; Weber, *Northern Railroads*; Black, *Railroad of the Confederacy*; Turner, *Victory Rode the Rails*; Gabel, *Railroad Generalship* and *Rails to Oblivion*; and chap. 1 of Hochfelder, *Telegraph in America*. For memoirs and biographies, see Haupt, *Reminiscences*; O'Brien, *Telegraphing in Battle*; *Autobiography of Andrew Carnegie*; and Kamm, "Civil War Career of Thomas A. Scott." Recent works of note include Clark, *Railroads in the Civil War*; Angevine, *Railroad and the State*; Wilson, *Business of Civil War*; and Army, *Engineering Victory*.

6. Tracks laid along surface streets eventually provided a means for transferring rail cars between the two rail lines, but the gap was not formally bridged until the 1870s when Pennsylvania Railroad affiliates built the Baltimore and Potomac Tunnel under West Baltimore and the Union Tunnel in East Baltimore.

7. *Autobiography of Andrew Carnegie*, 99–100. In his later years, Carnegie liked to boast that he was among the nation's first "defenders" to shed blood during the Civil War. See also O'Brien, *Telegraphing in Battle*, 10.

8. Kamm, "Civil War Career of Thomas A. Scott," 22.

9. Cameron and Thomson had a close business relationship. Before the outbreak of the war, the Pennsylvania Railroad had fought with the B&O for control of Cameron's Northern Central Railroad, which provided a strategic link between Harrisburg, Pennsylvania, and Baltimore. Cameron favored the Pennsylvania and used his controlling interest in the company to forge an alliance with Thomson and give the Pennsylvania Railroad a deep-water outlet on the Chesapeake Bay. Turner, *Victory Rode the Rails*, 47–48.

10. For a lively, firsthand account of the Baltimore rioting from a commercial telegrapher, see O'Brien, *Telegraphing in Battle*, 5–9.

11. Turner, *Victory Rode the Rails*, 25.

12. Turner, *Victory Rode the Rails*, 29–30.

13. Cameron deliberately excluded John Garrett, the B&O's president, from his correspondence with Thomson and Scott. He disliked Garrett personally and questioned his loyalty to the Union cause. Turner, *Victory Rode the Rails*, 50–52.

14. Turner, *Victory Rode the Rails*, 35.

15. Turner, *Victory Rode the Rails*, 33. For a detailed account of Scott's perilous journey from Harrisburg to Washington, see Kamm, "Civil War Career of Thomas A. Scott," 32–36.

16. Simon Cameron, Washington, DC, April 27, 1861, *ORA*, series 1, vol. 2, pt. 1, 603.

17. Kamm, "Civil War Career of Thomas A. Scott," 35; Turner, *Victory Rode the Rails*, 56.

18. Turner, *Victory Rode the Rails*, 58.

19. Lord, *Lincoln's Railroad Man*, 41.

20. O'Brien, *Telegraphing in Battle*, 21.

21. Kamm, "Civil War Career of Thomas A. Scott," 42.

22. Kamm, "Civil War Career of Thomas A. Scott," 41. The B&O main line from Baltimore to the Ohio River ran along the Virginia and Maryland border. Confederate raids and military occupations along the route periodically cut the line, particularly during the early years of the conflict, and made it unreliable as a transportation route.

23. Prior to the war, agents shipping goods from Baltimore and points north to Alexandria, Virginia, and to points south, had to cart their shipments through the streets of Washington, and across the Potomac to a railhead on the Virginia shore.

24. The current (ca. 1904) CSX Corporation Long Railroad Bridge across the Potomac River between Washington and Arlington, Virginia, lies a few hundred feet upstream from the site of the original Long Bridge.

25. Kamm, "Civil War Career of Thomas A. Scott," 43; Turner, *Victory Rode the Rails*, 59.

26. Kamm, "Civil War Career of Thomas A. Scott," 43.

27. Lord, *Lincoln's Railroad Man*, 39. Scott issued one of the first orders recognizing military railroads in July of 1861. He requested that federal military officers "furnish upon requisition from A. [Carnegie] superintendent in charge of railways, such facilities, rations, etc. as he may desire for the forces under his charge." Order of Thomas A. Scott, "General Manager, Government Railways and Telegraphs," quoted in Lord, 39.

28. R. F. Morley to Simon Cameron, October 3, 1861, *ORA*, series 3, vol. 1, pt. 1, 674.

29. Kamm, "Civil War Career of Thomas A. Scott," 43; Plum, *Military Telegraph*, 75.

30. Plum, *Military Telegraph*, 75.

31. The Battle of Bull Run, the first major engagement of the war in July of 1861, involved a fight for control of the junction of the Orange & Alexandria and Manassas Gap Railroads in northern Virginia.

32. Plum, *Military Telegraph*, 78.

33. Thomas A. Scott to Irvin McDowell, July 21, 1861, *ORA*, series 1, vol. 2, pt. 1, 750.

34. Kamm, "Civil War Career of Thomas A. Scott," 44–45.

35. Kamm, "Civil War Career of Thomas A. Scott," 46–48. Another aspirant for the position of assistant secretary of war was former Pennsylvania Railroad chief engineer Herman Haupt. Haupt, who would later play a dominant role in the management and operation of the USMRR, had originally been Scott's boss when he first went to work for the PRR in 1850. Cameron disliked Haupt for personal reasons and did not nominate him for the position.

36. Kamm, "Civil War Career of Thomas A. Scott," 48.

37. Thomas A. Scott to the Pennsylvania Railroad Board of Managers, August 21, 1861, 67, and Minutes of the Pennsylvania Railroad Board of Managers, September 25, 1861, 94. Accession 1807: Pennsylvania Railroad Corporation, Hagley Museum and Library, record group 1, series 6, vol. 4.

38. Plum, *Military Telegraph*, 79.

39. Kamm, "Civil War Career of Thomas A. Scott," 51.

40. See chap. 2 for more information on Stager's antebellum work for Western Union. Weber, *Northern Railroads*, 25.

41. Kamm, "Civil War Career of Thomas A. Scott," 50; Ellet, "Anson Stager."

42. Plum, *Military Telegraph*, 91.

43. Plum, *Military Telegraph*, 127.

44. Plum, *Military Telegraph*, 128.

45. Under Stager's plan, military telegraphers were treated as independent contractors, much like civilians working in other capacities for the Quartermaster's Department. Wilson, *Business of Civil War*, 2.

46. Plum, *Military Telegraph*, 130.

47. Plum, *Military Telegraph*, 131.

48. R. F. Morley to Simon Cameron, October 3, 1861, *ORA*, series 3, vol. 1, pt. 1, 673.

49. A reporter for the *New York Herald* noted that Scott had had "no such thing as a release for the duties of his office during the day, the night or on Sunday since the rebellion broke out." Kamm, "Civil War Career of Thomas A. Scott," 83.

50. Thomas A. Scott to Edwin M. Stanton, January 23, 1862, *ORA*, series 3, vol. 1, pt. 1, 808.

51. Angevine, *Railroad and the State*, 134–135. Also see US Congress, *Congressional Record*, 37th Cong., 2nd sess., 1862, chap. 15, p. 334.

52. Kamm, "Civil War Career of Thomas A. Scott," 85.

53. Daniel C. McCallum, "Superintendent's Report," from the *Erie Annual Report (1855)* reprinted in Chandler, *Railroads*, 102. See chap. 2 of this volume for more information on McCallum's work for the Erie Railroad.

54. Daniel C. McCallum, General Manager Military Railroads, to Montgomery C. Meigs, Quartermaster General, Washington, DC, April 7, 1862; Correspondence, Press Copies of Letters Sent by Gen. Daniel C. McCallum, February 1862–July 1866, 76–78; Records of the Office of the Director and General Manager, Military Railroads United States; Records of the Office of the Quartermaster General, record group 92, National Archives (hereafter OQG, RG92, National Archives).

55. Irvin McDowell to Edwin M. Stanton, April 10, 1862, *ORA*, series 1, vol. 12, pt. 3, 63.

56. Edwin M. Stanton to Daniel C. McCallum, April 12, 1862, *ORA*, series 1, vol. 12, pt. 3, 70.

57. Irvin McDowell to P. H. Watson, May 11, 1862, *ORA*, series 1, vol. 12, pt. 3, 169–170.

58. Haupt, *Reminiscences*, 44.

59. Lord, *Lincoln's Railroad Man*, 55.

60. Haupt, *Reminiscences*, 312–313.

61. Haupt, *Reminiscences*, 47–48.

62. Lord, *Lincoln's Railroad Man*, 77.

63. Haupt, *Reminiscences*, 312.

64. Haupt, *Reminiscences*, 143.

65. Haupt, *Reminiscences*, 173. Halleck's orders directly conflicted with McCallum's official status as "military director and superintendent of [government] railroads." Lord, *Lincoln's Railroad Man*, 42.

66. Lord, *Lincoln's Railroad Man*, 107.

67. Herman Haupt to Agents and Employees of the Military Railroad Department, December 19, 1862, *ORA*, series 3, vol. 2, pt. 1, 952.

68. Haupt blamed Stanton for the disorganized state of railroads in the West and claimed that the secretary of war had obstructed efforts to improve the situation earlier in the conflict. Letter from Herman Haupt to Benjamin Franklin Butler, October 1, 1863, in Butler, *Private and Official Correspondence*, 636.

69. Daniel C. McCallum to Edwin M. Stanton, January 19, 1864, *ORA*, series 1, vol. 32, pt. 2, 145.

70. Daniel C. McCallum to Edwin M. Stanton, November 27, 1864, *ORA*, series 3, vol. 4, pt. 1, 948.

71. Daniel C. McCallum to Edwin M. Stanton, January 19, 1864, *ORA*, series 1, vol. 32, pt. 2, 145.

72. Entry for November 19, 1863, Journal of Events Maintained at Alexandria, Virginia, 1863–1864; Records of the Office of the Director and General Manager, Military Railroads United States; OQG, RG92, National Archives.

73. Entry for November 20, 1863, Journal of Events Maintained at Alexandria, Virginia, 1863–1864; Records of the Office of the Director and General Manager, Military Railroads United States; OQG, RG92, National Archives.

74. Entry for December 11, 1863, Journal of Events Maintained at Alexandria, Virginia, 1863–1864; Records of the Office of the Director and General Manager, Military Railroads United States; OQG, RG92, National Archives.

75. Report of Gen. Daniel C. McCallum, Director and General Manager Military Railroads, United States, for the Year Ending June 30, 1864, November 27, 1864, *ORA*, series 3, vol. 4, pt. 1, 953.

76. Daniel C. McCallum to Anson Stager, Washington, DC, October 14, 1862; Correspondence, Telegrams Sent by General Daniel C. McCallum, 1862–1864; Records of the Office of the Director and General Manager, Military Railroads United States; OQG, RG92, National Archives.

77. Lord, *Lincoln's Railroad Man*, 116.

78. Lord, *Lincoln's Railroad Man*, 116.

79. Lord, *Lincoln's Railroad Man*, 249–252; letter from Herman Haupt to Benjamin F. Butler, October 1, 1863, in Butler, *Private and Official Correspondence*, 636. Haupt's biographer Francis Lord speculates that McCallum may have also played a role in Haupt's dismissal, since he was a close acquaintance of Governor John Andrew. Andrew eventually succeeded in removing Haupt from the Hoosac Tunnel project. Lord, *Lincoln's Railroad Man*, 245, 255–257.

80. Lord, *Lincoln's Railroad Man*, 281.

81. Lord, *Lincoln's Railroad Man*, 281–282.

82. Annual Report of Daniel C. McCallum, May 26, 1866, *ORA*, series 3, vol. 5, no. 1, 978.

83. Lord, *Lincoln's Railroad Man*, 270–271. McCallum's wartime efforts eventually won him a promotion to brigadier general in September of 1864. At the end of the war, Congress granted McCallum a brevet appointment as major general. Lord, *Lincoln's Railroad Man*, 246, 257.

84. Weber, *Northern Railroads*, 220–221. In 1862, the federal government negotiated a troop rate with major civilian railroads of two cents per mile for each soldier and a freight rate 10 percent lower than the comparable civilian rate between two shipping points for bulk goods, animals, and other types of cargo. Weber, *Northern Railroads*, 130.

85. Weber, *Northern Railroads*, 44, 50, 55, 58.

86. Weber, *Northern Railroads*, 63, 88.

87. Weber, *Northern Railroads*, 44.

88. *Autobiography of Andrew Carnegie*, 93. Carnegie believed he had hired the first night dispatchers in the United States. It is impossible to verify his claim, but he almost certainly hired the Pennsylvania Railroad's first night dispatcher.

89. Letter from G. A. Nicolls to J. O. Stearns, July 21, 1863. Accession 1520: Philadelphia & Reading Railroad Company, series 1, box 111-1, Hagley Museum and Library (hereafter Reading Collection).

90. Letter from G. A. Nicolls to J. O. Stearns, September 10, 1864. Reading Collection, series 1, box 111-1. Underline in original document.

91. See chap. 2 for the history of the Illinois Central Railroad's relationship to Caton's Illinois Central Telegraph Company.

92. Letter from William Osborn to J. D. Caton, February 20, 1863. Illinois Central Railroad Collection, IC1/06.1/vol. 5, Newberry Library. Quoted in Yates, *Control through Communication*, 106.

93. Letter from Osborn to Caton, December 22, 1863. Illinois Central Railroad Collection, IC1/06.1/vol. 5. Quoted in Yates, *Control through Communication*, 106. In her analysis of the letter, Yates concludes that "the language of this statement clearly implies that the situation was a recent one." Yates, 106.

94. Usselman, *Regulating Railroad Innovation*, 126.

95. Contract between the Illinois & Mississippi Telegraph Company and the Chicago, Burlington & Quincy Railroad, April 15, 1863. Chicago, Burlington & Quincy Railroad Papers, CB&Q Abstract of Telegraph Contracts, 33.1870.5.1 (Contract File, 1856–1881), Newberry Library.

96. The term *American system* was typically used to distinguish this set of practices from other train management methodologies such as the "British system." A correspondent to the *Railroad Gazette* noted, for instance, that a British railroad engineer "had organized one of the Indian railways on the American system" instead of the British system. See "American Train Dispatching," *Railroad Gazette*, February 17, 1872.

97. "The Chicago, Burlington & Quincy Railroad Telegraph," *Railroad Gazette*, June 17, 1871.

98. Shaw, *History of Railroad Accidents*, 120.

99. Shaw, *History of Railroad Accidents*, 71–72. At the time, the Erie Railroad was operating under the guidelines set forth in its 1857 *Rulebook*. Under the provisions of the rule book, movement of trains by telegraph was considered to be an exceptional practice. The rules for running trains by telegraph had not changed since the mid-1850s when McCallum developed them. See chap. 2 for more details.

100. Usselman, *Regulating Railroad Innovation*, 126.

101. Annual Report of Daniel C. McCallum, May 26, 1866, *ORA*, series 3, vol. 5, no. 1, 1003; Plum, *Military Telegraph*, 380.

102. Chandler, *Visible Hand*, 156.

103. Annual Report of Col. Anson Stager, September 15, 1865, *ORA*, series 3, vol. 5, no. 1, 361.

104. Plum, *Military Telegraph*, 337. Plum estimated the average cost of each telegram for the federal government at 40 cents, or roughly $5.17 in 2017 dollars. Inflation calculation based on Consumer Price Index data from Measuring Worth, http://www.measuringworth.com/uscompare (accessed July 6, 2018).

105. Annual Report of Col. Anson Stager, September 15, 1865, *ORA*, series 3, vol. 5, no. 1, 359.

106. Lord, *Lincoln's Railroad Man*, 270.

107. Annual Report of Daniel C. McCallum, May 26, 1866, *ORA*, series 3, vol. 5, no. 1, 1000.

108. Annual Report of Col. Anson Stager, September 15, 1865, *ORA*, series 3, vol. 5, no. 1, 359.

Chapter 4 • The American System

1. Field, "Magnetic Telegraph, 408; Board of Railroad Commissioners, *Third Annual Report*, 169–170. In the early 1870s, construction costs for American railroads averaged approximately $50,000 per mile. By comparison, construction and operating costs for telegraph lines averaged $100-$200 per mile.

2. Telegraph-based block signals in Britain became widespread during the 1860s and 1870s. Conceptually, they were based on British telegraph pioneer William F. Cooke's ideas about railroad management that he laid out in his 1842 pamphlet *Telegraphic Railways*. See chap. 1. Manually controlled signals would indicate to train drivers when it was safe to enter a stretch of track. Once the train entered the block, a signalman would use the telegraph to convey this information to the next signalman, who would then notify the first signalman when it was safe to reopen the block. After a number of severe railroad accidents in the 1880s, Parliament mandated in 1889 that all British railroads employ absolute block signals on their lines. Solomon, *Railroad Signaling*, 32. Other European nations used variations on block signaling during the nineteenth century.

3. This chapter focuses primarily on eastern and midwestern railroads. The western transcontinental railroads of the postwar era, such as the Union Pacific and later a host of other lines, both functional and speculative, often operated according to their own internal logic and the profit motives of the founders. As Richard White has noted, "Examining corporations through nineteenth-century western railroads is like looking at them through a funhouse mirror." Western railroads utilized the telegraph for operational management, but in a less rigorous and organized manner than many of their eastern and midwestern peers. White, *Railroaded*, 234.

4. See chaps. 2 and 3.

5. Turner and Jacobus, *Connecticut Railroads*, 4.

6. Adams, *Notes on Railroad Accidents*, 128.

7. Adams, *Notes on Railroad Accidents*, 133; Shaw, *History of Railroad Accidents*, 78.

8. Adams, *Notes on Railroad Accidents*, 129.

9. Shaw, *History of Railroad Accidents*, 78.

10. Adams, *Notes on Railroad Accidents*, 133.

11. Adams, *Notes on Railroad Accidents*, 132–135.

12. Adams, *Notes on Railroad Accidents*, 137–140. For an analysis of "chain reaction" accidents and complex technical systems, see Perrow, *Normal Accidents*.

13. "The Revere Accident," from the *Telegrapher*, reprinted in *Railroad Gazette*, September 16, 1871.

14. Usselman, *Regulating Railroad Innovation*, 120.

15. "The Massachusetts Railroad Commissioners on the Revere Disaster," *Railroad Gazette*, September 23, 1871.

16. Board of Railroad Commissioners, *Third Annual Report*, 129.

17. Board of Railroad Commissioners, *Third Annual Report*, 129–130.

18. Board of Railroad Commissioners, *Third Annual Report*, 133.

19. Board of Railroad Commissioners, *Third Annual Report*, 132–136.

20. Board of Railroad Commissioners, *Third Annual Report*, 137.

21. This practice came to be known as manual telegraphic block signaling.

22. Board of Railroad Commissioners, *Third Annual Report*, 137.

23. Board of Railroad Commissioners, *Third Annual Report*, 137–138.

24. Board of Railroad Commissioners, *Third Annual Report*, 138.

25. Board of Railroad Commissioners, *Third Annual Report*, 138.

26. Adams, *Notes on Railroad Accidents*, 65.

27. Usselman, *Regulating Railroad Innovation*, 126; Weber, *Northern Railroads*, 43–45.

28. Usselman, *Regulating Railroad Innovation*, 120–121.

29. The Reading Railroad was an early adopter of block signaling. The railroad had instituted a form of manual signaling along its main line between Reading and Philadelphia in the mid-1860s. The Reading utilized timed signals, rather than telegraphy, to maintain safe intervals between trains. Usselman, *Regulating Railroad Innovation*, 124–125.

30. Camden & Amboy Railroad and Transportation Company, Minutes of the Executive Committee and Joint Board of the Camden & Amboy and New Jersey Railroad and Transportation Company, March 27, 1865, 65. Accession 1807: Pennsylvania Railroad Corporation, record group 3: Lines East Companies, Camden & Amboy Railroad and Transportation Company, minute book 6, Hagley Museum and Library.

31. Calvert, "Notes on Pennsylvania Railroad Operation and Signaling."

32. For more information on early British telegraph technologies, see chap. 1. Canadian railroads used train management practices similar to those of US railroads. French and German firms based their practices on the British model. Indian colonial railroads, with a few exceptions, also based their operating practices on the British model.

33. Board of Railroad Commissioners, *Third Annual Report*, 150–151.

34. Ashbel Welch, "Report on Railroad Safety Signals to the Railroad Convention, Held at the St. Nicholas Hotel, New York, October 17, 1866," *Engineering News*, February 11, 1882, 45.

35. Pennsylvania Railroad Corporation, Minutes of the Board of Managers (hereafter PRR Board Minutes), April 25, 1871, 459. Accession 1807: Pennsylvania Railroad Corporation, record group 1, series 6, vol. 5, Hagley Museum and Library.

36. Adams, *Notes on Railroad Accidents*, 276–277.

37. Board of Railroad Commissioners, *Third Annual Report*, 169–170. Adjusted for inflation, this would be over $376,000 in 2017 dollars for equipment and $37,600 in monthly wages. Inflation calculation based on Consumer Price Index data from Measuring Worth, http://www.measuringworth.com/uscompare (accessed July 6, 2018).

38. By 1871, the Pennsylvania Railroad's capitalization was nearly $400,000,000. This was 13 percent of the total capital invested in the American railroad industry. Chandler, *Visible Hand*, 151.

39. Sixteen percent of the railroad companies located in the mid-Atlantic began defaulting on their bond payments by the end of 1873. In the Midwest, 34 percent of the railroads defaulted on their investments during the same year. Many of these lines declared bankruptcy and went into receivership for the rest of the decade. Stover, "Southern Railroad Receivership," 41n8.

40. The legality of railroad-owned commercial telegraph lines remained ambiguous until 1879 when Congress approved the Butler Amendment to the Army Appropriation Bill. The new law explicitly permitted railroads to operate interstate telegraph networks for commercial use if they agreed to federal rate regulations. "The Army Appropriation Bill Passed," *New York Times*, February 9, 1879. For more information about the history of the Butler Amendment, see John, *Network Nation*, 164–166, 172; Wolff, *Western Union*, 250–253. Also see HR178 (50-1), January 31, 1888, "Military and Postal Telegraph," 6; Pennsylvania Railroad, PRR Board Minutes, vol. 8, July 23, 1879, 319. Accession 1807: Pennsylvania Railroad Corporation, record group 1, series 6, Hagley Museum and Library.

41. In 1879, WU president Norvin Green asserted that almost all American railroads depended on Western Union for telegraph service (he put the ratio at nineteen out of twenty). He argued that only a small number of railroads, a "half a dozen lines" nationwide, possessed

their own telegraph infrastructure and operated their own telegraph networks. *Hearing before the Committee on Railroads*, Senate, 45th Cong. 2 (1879).

42. Thompson, *Wiring a Continent*, 422, 424, 426.

43. Wolff, *Western Union*, 247.

44. "Railroad Telegraphy," *Telegrapher*, September 14, 1867.

45. Hochfelder, *Telegraph in America*, 57.

46. Wolff, *Western Union*, 131–133. Under a typical arrangement with Western Union, railroads provided telegraph poles and labor to erect them, while WU contributed wires and insulators and an employee to superintend construction of the line. WU also provided telegraph instruments. Railroads typically granted free passes to WU employees on official business and free transportation of telegraph supplies for line construction and maintenance. Many of these contractual arrangements were challenged by either railroad firms or WU in state and federal courts over the course of the nineteenth century, and a significant body of case law emerged regarding the rights and contractual obligations of both parties. See Cooke, *Treatise on Telegraph Law*, 29–66.

47. "The Revere Accident," from the *Telegrapher*, reprinted in *Railroad Gazette*, September 16, 1871.

48. Editorial, *Telegrapher*, September 14, 1872.

49. Editorial, *Telegrapher*, September 14, 1872.

50. "Railroad Wires. Their Uses and Their Abuses," *Journal of the Telegraph*, December 15, 1871.

51. *Journal of the Telegraph*, January 15, 1872, and February 15, 1872.

52. PRR Board Minutes, vol. 4, May 3, 1865, 450. Accession 1807: Pennsylvania Railroad Corporation, record group 1, series 6, Hagley Museum and Library.

53. Wolff, *Western Union*, 131–133.

54. The P&A and the A&P were separate commercial entities, despite their similar-sounding names. The P&A was chartered in Pennsylvania, and the A&P was chartered in New York. Wolff, *Western Union*, 129–131.

55. John, *Network Nation*, 158–159.

56. PRR Board Minutes, vol. 7, September 22, 1875, 73; November 10, 1875, 92; November 24, 1875, 99. Accession 1807: Pennsylvania Railroad Corporation, record group 1, series 6, Hagley Museum and Library.

57. John, *Network Nation*, 160. Also see Wolff, *Western Union*, 209–231, for a detailed discussion and analysis of Gould's first campaign against Western Union using the A&P Telegraph Company.

58. John, *Network Nation*, 116; Wolff, *Western Union*, 250–251.

59. PRR Board Minutes, vol. 8, July 23, 1879, 319. Accession 1807: Pennsylvania Railroad Corporation, record group 1, series 6, Hagley Museum and Library.

60. PRR Board Minutes, vol. 8, January 17, 1880, 436–438. Accession 1807: Pennsylvania Railroad Corporation, record group 1, series 6, Hagley Museum and Library.

61. PRR Board Minutes, vol. 9, September 22, 1880, 85; October 13, 1880, 92. Accession 1807: Pennsylvania Railroad Corporation, record group 1, series 6, Hagley Museum and Library.

62. John, *Network Nation*, 166–167.

63. "The War of the Telegraphs," *New York Times*, April 5, 1880; "National Capital Topics," *New York Times*, December 15, 1886. As the lawsuits wound their way through the nation's

courts, Gould eventually found himself on the opposite side of the legal struggle once he took control of Western Union in 1881 and gave up managerial control of the Union Pacific in 1883. For a more in-depth study of Gould's business activities and his corporate legacy, see Klein, *Life and Legend of Jay Gould*.

64. John, *Network Nation*, 170. For more on Gould's acquisition and control of Western Union, see Grodinsky, *Jay Gould*, chaps. 14 and 23.

65. When the PRR contract expired in 1901, the railroad once again prepared for battle against Western Union, now under the control of Jay Gould's son, George. Like his father, George Gould was a network builder, and around 1900 he began trying to build an eastern extension of the Wabash Railroad into Pittsburgh to draw local traffic away from the Pennsylvania Railroad and onto his railroad network. The PRR fought the move vigorously, and after failing to stop Gould's progress through other means, PRR president Alexander J. Cassatt kicked WU staff out of all telegraph offices along the PRR right-of-way. When this action failed to change Gould's mind, Cassatt gave WU until the end of November 1902 to remove all its wires and poles from the PRR lines east of Pittsburgh. WU obtained a court injunction until the spring of 1903, but as soon as it expired, ten thousand PRR workers tore down all WU telegraph poles and more than twenty-five thousand miles of wire. This rather drastic action failed to dissuade Gould, who completed the rail line in 1904 after John D. Rockefeller stepped in to mediate a truce between the warring parties. As part of the gentlemen's agreement, Cassatt compensated WU for the value of its destroyed property, estimated at around $500,000. George Gould's corporate empire eventually collapsed following the Panic of 1907, and he lost control of WU by 1912. PRR Board Minutes, vol. 9, November 1, 1881, 346–353. Accession 1807: Pennsylvania Railroad Corporation, record group 1, series 6, Hagley Museum and Library; Campbell, "The Wabash—The Gould Downfall." Also see Churella, *Building an Empire*, 655–656; Hochfelder, *Telegraph in America*, 40.

66. Determining the exact number of railroad accidents during the nineteenth century is exceedingly difficult, as railroad accident expert Robert B. Shaw notes. Until the 1880s, daily newspapers were the major source of information on collisions and other types of railroad accidents. Newspapers tended to focus on the most sensational accidents and largely ignored minor collisions. Beginning in the 1870s, the railroad trade press began to report on railroad accidents in a more systematic manner. Similarly, state railroad commissions began collecting data on accidents during this era as part of the efforts to advocate for industry reforms. After its creation in 1901, the Interstate Commerce Commission began gathering reports of accidents from railroads across the nation and published its first *Accident Bulletin* in 1902. Shaw, *History of Railroad Accidents*, 6–10. Economist Mark Aldrich conducted a detailed survey of railroad accidents in the nineteenth century. According to his research, roughly 1,260 collisions occurred between the mid-1870s and mid-1880s, of which approximately 60 percent were rear-end collisions. During the same time period, 651 passenger fatalities are recorded, many of which resulted from these collisions. Aldrich, *Death Rode the Rails*, 311–312, 319.

67. Usselman, *Regulating Railroad Innovation*, 73.

68. "Notes on the Management and Discipline of American Railroads," *Railroad Gazette*, February 3, 1872.

69. "Notes on the Management and Discipline of American Railroads," *Railroad Gazette*, February 3, 1872.

70. "X Replies," *Railroad Gazette*, February 17, 1872.

71. "America Replies," *Railroad Gazette*, February 17, 1872.

72. "Hindoo Responds," *Railroad Gazette*, February 24, 1872. Also see Aldrich, *Death Rode the Rails*, 319.

73. "Hindoo Responds," *Railroad Gazette*, February 24, 1872.

74. "Train Dispatching," *Railroad Gazette*, April 13, 1872.

75. "Hindoo Responds," *Railroad Gazette*, March 2, 1872. Hindoo cited eight head-on collisions that had taken place in the month of February 1872 alone.

76. "X Responds," *Railroad Gazette*, April 20, 1872.

77. "J. H McNairn Responds to Hindoo," *Railroad Gazette*, April 6, 1872.

78. "Observer Responds to Hindoo," *Railroad Gazette*, April 27, 1872.

79. "Hindoo Responds to His Critics," *Railroad Gazette*, May 18, 1872.

80. "From the *Telegrapher*," *Railroad Gazette*, March 16, 1872.

81. "Dispatch of W. W. Wells," *Railroad Gazette*, April 13, 1872.

82. "Railway Association of America," *Railroad Gazette*, May 31, 1873.

83. "Railway Association of America," *Railroad Gazette*, May 31, 1873. Between 1873 and 1877, an average of six trains collided each month on American rail lines. British monthly totals were significantly lower. Mark Aldrich has calculated that, ironically, passenger fatality rates per miles traveled on British railroads were significantly higher than on American railroads in the 1870s. In 1873, an average of 19.21 American passengers were killed in accidents, while an average of 37.96 British passengers died in train accidents. This statistic reflects the greater mileage and lower ridership of American railroads compared to British lines, rather than safer operating practices on American rail lines. Aldrich, *Death Rode the Rails*, 311, 315, 319.

84. "Railway Association of America," *Railroad Gazette*, May 31, 1873. Adams and other railroad reformers drew attention to air brakes as a possible remedy for train accidents, but there was little enthusiasm for them among the railroad community. A bill mandating air brakes for passenger trains was introduced in Congress in 1873, but the legislation went nowhere. It would take until 1900 for air brakes to become a required safety appliance for trains in the United States. Usselman, *Regulating Railroad Innovation*, 120–123.

Chapter 5 • The Struggle for Standards

1. For example, see "Train Dispatching," *Railroad Gazette*, January 19, 1877.

2. The debate over the development of the rules and their implementation fills the first published volume of the GTC's proceedings. See *Proceedings of the General Time Convention and Its Successor the American Railway Association from Its Organization April 14, 1886 to October 11, 1893 Inclusive*, vol. 1 (New York: The American Railway Association, 1893) (hereafter *GTC*). For more information on industrial standards setting in the nineteenth and twentieth centuries, see Russell, "'Industrial Legislatures'" and Russell, *Open Standards and the Digital Age: History, Ideology, and Networks.*

3. Usselman, *Regulating Railroad Innovation*, 64–65.

4. See chap. 1 for more information on early standards setting in New England.

5. Dunlavy, "Organizing Railroad Interests," 137. Also see Usselman, *Regulating Railroad Innovation*, 114.

6. Angevine, *Railroad and the State*, 134–135. Also see US Congress, *Congressional Record*, 37th Cong., 2nd sess., 1862, Chapter 15, p. 334.

7. Dunlavy, "Organizing Railroad Interests," 137–139.

8. Ashbel Welch, "Report on Railroad Safety Signals to the Railroad Convention, Held at the St. Nicholas Hotel, New York, Oct. 17, 1866," *Engineering News*, February 11, 1882.

9. Ashbel Welch, "Report on Railroad Safety Signals to the Railroad Convention, Held at the St. Nicholas Hotel, New York, Oct. 17, 1866," *Engineering News*, February 11, 1882.

10. Dunlavy, "Organizing Railroad Interests," 138.

11. "Uniformity in Codes of Rules," *Railroad Gazette*, January 17, 1884.

12. Chandler, *Visible Hand*, 153–155.

13. For more information on the PRR's expansion during this era, see Churella, *Building an Empire*, chaps. 11–13.

14. Chandler, *Visible Hand*, 155–156.

15. Usselman, *Regulating Railroad Innovation*, 189.

16. Usselman, *Regulating Railroad Innovation*, 189–190.

17. The ATO's name emphasized the fact that it was an organization composed of *the* transportation officers from the PRR, rather than simply an association of railroad transportation officials.

18. "Constitution of the Association of the Transportation Officers," May 20, 1879, quoted in Usselman, *Regulating Railroad Innovation*, 190.

19. Usselman, *Regulating Railroad Innovation*, 190. For examples of other questions see "Minutes of the Association of the Transportation Officers of the Pennsylvania Railroad," January 14, 1880, 11–12. Accession 1810: Papers of the Association of the Transportation Officers of the Pennsylvania Railroad (henceforth PRR ATO Files), box 407, folder 2, Hagley Museum and Library.

20. *Rules and Regulations for the Government of the Transportation Department of the Pennsylvania Railroad* (1874), 40–44, 73–78. The 1874 rule book also addressed the PRR's 1871 acquisition of the Camden & Amboy Railroad by including rules for moving trains between Philadelphia and Jersey City using the sophisticated telegraphic block signaling arrangement developed by Ashbel Welch in the mid-1860s (discussed in chap. 4). Pennsylvania Railroad Corporation, Minutes of the Board of Managers, April 25, 1871, 459. Accession 1807: Pennsylvania Railroad Corporation, record group 1, series 6, vol. 5, Hagley Museum and Library; Usselman, *Regulating Railroad Innovation*, 125.

21. Anderson, *Train Wire*, ix; J. A. Anderson, "Rules for the Movement of Trains by Telegraphic Train Orders," *Railroad Gazette*, June 14, 1873.

22. Anderson, *Train Wire*, vi.

23. *Rules and Regulations for the Government of the Transportation Department of the Pennsylvania Railroad* (1874), 32; *Rules and Regulations for the Government of the Transportation Department of the Pennsylvania Railroad* (1882), 74.

24. *GTC*, 139. Anecdotal evidence does indicate that most PRR dispatchers had begun using double-order dispatching by the beginning of the 1880s.

25. "The Train Wire. A Discussion on Train Dispatching: Train Dispatching," *Railway Age*, October 5, 1882.

26. "The Train Wire. A Discussion on Train Dispatching: Chapter 1 Continued," *Railway Age*, October 12, 1882; "The Train Wire. A Discussion on Train Dispatching: The Operator," *Railway Age*, October 19, 1882.

27. Anderson's book received highly favorable reviews in the *Railroad Gazette* the following year. "A Work on Train Dispatching," *Railroad Gazette*, October 24, 1884.

28. Usselman, *Regulating Railroad Innovation*, 114–115, 291–292.

29. Usselman, *Regulating Railroad Innovation*, 273–274. For more information about public safety campaigns in response to collisions in the 1870s, see Usselman, *Regulating Railroad Innovation*, 120–123.

30. Usselman, *Regulating Railroad Innovation*, 274.

31. Aldrich, *Death Rode the Rails*, 319; "Uniformity in Codes of Rules," *Railroad Gazette*, January 17, 1884.

32. Dunlavy, "Organizing Railroad Interests," 139.

33. McCrossen, *Marking Modern Times*, 94.

34. Bartky, *Selling the True Time*, 138. For a detailed study of the GTC's efforts to establish standardized time zones for the American railroad industry and the broader national and international context of the effort, see chap. 11 of Bartky's study. For more information about railroad timekeeping practices and technologies and the role of Western Union as a provider of timekeeping services to railroads, see McCrossen, *Marking Modern Times*, chap. 4.

35. William F. Allen, "Standard Railway Time," *Century* 26, no. 5 (1883): 797, quoted in Pietruska, *Looking Forward*, 13.

36. Bartky, *Selling the True Time*, 139.

37. *GTC*, 692.

38. *GTC*, 703; Bartky, *Selling the True Time*, 141. National civic adoption of the railroad time zones would follow within the next few years.

39. *GTC*, 692. Train-crew-operated signals included locomotive whistles operated by the engineer and various colored flags and lights mounted on the front and rear of locomotives and cabooses. Whistle codes allowed engineers to communicate with conductors and brakemen on their trains. Flag and light signals indicated whether a particular train was operating on its own schedule, as an unscheduled "wild" train, or as a part of a multitrain convoy.

40. *GTC*, 698.

41. *GTC*, 710.

42. *GTC*, 704.

43. *GTC*, 705.

44. *GTC*, 706.

45. "Uniform Signals—A Railroad Bible," *Railway Age*, July 24, 1884.

46. *GTC*, 717.

47. *GTC*, 718.

48. Talbott and Hobart, *Biographical Directory*, 189, 252; PRR ATO Files, January 14, 1880, box 407, folder 2, 8–9. *GTC*, 86.

49. *GTC*, 736.

50. *GTC*, 16

51. *GTC*, 48.

52. *GTC*, 86

53. *GTC*, 95.

54. *GTC*, 100, 139.

55. *GTC*, 139–140.

56. *GTC*, 199, 272.

57. Anderson, *Train Wire*, vi.

58. Anderson, *Train Wire*, viii. Seventy thousand miles of railroad track was just under half of the total mileage of railroad track in the United States in the 1890s. Corporate

coordination of this magnitude represented a remarkable achievement for the era. Field, "Magnetic Telegraph," 402.

59. Usselman, *Regulating Railroad Innovation*, 273–274. As Usselman notes, less than 15 percent of railroad track in the United States was operated by block signals or automatic signals in 1900. Also see Usselman, *Regulating Railroad Innovation*, 325–326.

Chapter 6 • Telegraphers and Regulators

1. "The Block System on Single Track Railroads," *Railroad Gazette*, October 27, 1893.

2. W. W. Ryder, "Dispatching Trains by Telephone," *Signal Engineer* 1, no. 2 (1908): 55–56; Usselman, *Regulating Railroad Innovation*, 325.

3. Interstate Commerce Commission officials estimated an average of thirteen telegraphers and dispatchers for every one hundred miles of rail line across the nation. *Eighth Annual Report*, 35.

4. Shaw, *History of Railroad Accidents*, 144; Aldrich, *Death Rode the Rails*, 71; McIsaac, *Order of Railroad Telegraphers*, 6–7.

5. Aldrich, *Death Rode the Rails*, 88. Only brakemen had a shorter average period of employment by railroad companies at 3.8 months, which is understandable, given the hazardous and physically demanding nature of their work. See Jepsen, *My Sisters Telegraphic*, for a detailed study of the lives of female telegraphers in both commercial and railroad service during the nineteenth century. Gabler, *The American Telegrapher*, offers a broader perspective on the work lives of telegraphers during this era.

6. W. W. Ryder, "Dispatching Trains by Telephone," *Signal Engineer* 1, no. 2 (1908): 55.

7. Estes, Federal Writers' Project Interview. The Interstate Commerce Commission estimated that the average daily wage for American telegraphers and dispatchers in 1892 was $1.93 (approximately $48.60 a day in 2017 dollars). Over the next ten years, wages for telegraphers remained stagnant, increasing slightly to $2.01 per day by 1902 (approximately $47.50 per day in 2017 dollars). In 1910, the average daily wage stood at $2.33 (approximately $45.20 per day in 2017 dollars), and by 1916 the daily wage had fallen to $2.21, assuming telegraphers worked a nine-hour day (approximately $35 per day in 2017 dollars). Since the annual rate of inflation averaged 1.34 percent over the same twenty-five-year period, telegraphers' buying power would have decreased significantly. By comparison, average daily wages for locomotive engineers rose from $3.68 to $4.55 between 1892 and 1910. When adjusted for inflation, however, their real income declined slightly during this period from approximately $92.70 to $88.20 in 2017 dollars. ICC, *Thirteenth Annual Report*, 40; ICC, *Twenty-Third Annual Report*, 38; ICC, *Twenty-Ninth Annual Report*, 28; Consumer Price Index data, MeasuringWorth.com, http://www.measuringworth.com/uscompare (accessed July 6, 2018).

8. Editorial, "Some of the Troubles of Operators," *Railroad Gazette*, January 29, 1892.

9. "Argument of a West Shore Telegraph Operator," *Railroad Gazette*, January 29, 1892.

10. Editorial, "Some of the Troubles of Operators," *Railroad Gazette*, January 29, 1892.

11. Editorial, "Some of the Troubles of Operators," *Railroad Gazette*, January 29, 1892.

12. McIsaac, *Order of Railroad Telegraphers*, 7, 222–223.

13. "Telegraphers' Strike on the Milwaukee & St. Paul," *Railway Age*, January 17, 1891; "The O.R.T. and the Rock Island Road," *Railway Age*, December 16, 1892; McIsaac, *Order of Railroad Telegraphers*, 14–15, 19.

14. J. C. Brown, "The Railroad Telegraph Department," *Railroad Gazette*, March 4, 1904; Editorial, *Railroad Gazette*, March 4, 1904.

15. "Train Dispatching," *Railway Age*, May 11, 1882.
16. "The Train Dispatcher," *Railway Age*, May 10, 1883.
17. "Double-Order Train Dispatching on the Chicago & Alton Railroad," *Railway Age*, August 9, 1883.
18. J. F. Mackie, "The Train Dispatchers' Attitude toward Their Companies—What They Ask," *Railway Age*, June 24, 1892.
19. "Train Dispatchers—Official or Employee," *Railway Age*, July 15, 1892; "About the New Order," *Railway Age*, February 3, 1893; "A List with a Lesson," *Railway Age*, February 3, 1894.
20. "His Life Ruined by a Cypher," *Journal of the Telegraph*, July 20, 1890. For a broader discussion of the telegraph in railroad fiction, see Stilgoe, "Sounders and Silence: Some Isolated Train-Order Stations in Fiction," 45–54.
21. McDonald, *Federal Railroad Safety Program*, 8.
22. Usselman, *Regulating Railroad Innovation*, 292.
23. Roosevelt, "Third Annual Message"; ICC, *Eighteenth Annual Report*, 105.
24. Roosevelt, "Third Annual Message"; McDonald, *Federal Railroad Safety Program*, 13.
25. See *H.R. 4438, H.R. 16676, and H.R. 18671: To Limit the Hours of Service of Railroad Employees: Hearing before Committee of Interstate and Foreign Commerce, House of Representatives*, 59th Cong. (1906).
26. "Limits to Trainmen's Hours," *Washington Post*, May 30, 1906.
27. "Limits to Trainmen's Hours," *Washington Post*, May 30, 1906; "Railway Affairs at the National Capitol. From Our Washington Correspondent," *Railway Age*, July 6, 1906.
28. "An Old-Fashioned Idea," *Railway Age*, December 25, 1906.
29. "For Service Pensions," *Washington Post*, January 10, 1907.
30. "Railroad Men Angry," *Washington Post*, February 17, 1907; "Hepburn Rewrites La Follette Bill," *New York Times*, February 17, 1907.
31. "Amend Bill for Trainmen," *Washington Post*, March 1, 1907; "Secretary Mosely Blamed," *Washington Post*, March 1, 1907; "Hours of Labor Cut," *Washington Post*, March 2, 1907.
32. "20,000 Messages Scare Congress," *New York Times*, March 2, 1907.
33. "Agree on Service Bill," *Washington Post*, March 4, 1907; "Act Limiting Hours of Railway Employees," *Railway Age*, March 8, 1907.
34. "Agree on Service Bill," *Washington Post*, March 4, 1907; "Act Limiting Hours of Railway Employees," *Railway Age*, March 8, 1907.
35. Editorial, "New Railroad Legislation" and "Hours of Labor Law," *Railroad Gazette*, March 8, 1907.
36. ICC, *Twenty-Second Annual Report*, 49; "Notes," *Railroad Gazette*, February 7, 1908.
37. Editorial, *Railroad Gazette*, February 21, 1908; "Miscellaneous," *Railroad Gazette*, March 6, 1908; "Railroads Want Hearing," *Washington Post*, February 13, 1908.
38. ICC, *Twenty-Second Annual Report*, 49; "Railroads Refused Time," *Washington Post*, March 3, 1908; "Railroad Men Confer," *Washington Post*, March 3, 1908.
39. ICC, *Twenty-Third Annual Report*, 52; ICC, *Twenty-Fourth Annual Report*, 23; ICC, *Twenty-Fifth Annual Report*, 83.
40. "Railroads Refused Time," *Washington Post*, March 3, 1908; "Notes," *Railroad Gazette*, January 31, 1908.

41. Editorial, "The Telegraphone in Train Dispatching," *Railway Age*, September 27, 1907; "Notes," *Railroad Gazette*, October 25, 1907; Editorial, "Pennsylvania Railroad Telegraph School," *Railroad Gazette*, September 20, 1907.

42. "General News Section," *Railway Age*, August 28, 1908.

43. AT&T acquired control of Western Union in 1909 and thus became involved directly with the commercial telegraph industry and indirectly with the railroad industry because of WU's right-of-way agreements with many railroads. AT&T eventually divested Western Union in 1913 as part of the Kingsbury Commitment to avoid an antitrust prosecution from the federal government. For more context on the 1909 acquisition and 1913 divestiture, see John, *Network Nation*, chaps. 7 and 10; and Hochfelder, *Telegraph in America*, chap. 5.

44. By 1907, the initial Bell telephone patents of the 1870s had expired or been voided by the courts, and a small but significant independent telephone industry emerged in the United States that attempted to coexist with the much larger Bell System run by American Telephone & Telegraph (AT&T). Railroads appear to have sourced telephone equipment for dispatching trials from independent equipment manufacturers, such as the Automatic Electric Company and the Kellogg Switchboard and Supply Co., and from Western Electric, the AT&T-controlled equipment manufacturer in Chicago. Western Electric became the dominant supplier of railroad telephone equipment in the post–World War I era. For more information on post–World War I railroad telephony, see Wills, "Telephone Train Dispatching."

45. Editorial, "Train Dispatching by Telephone on the Burlington," *Railway Age*, June 26, 1908.

46. Editorial, "Growth of Telephone Train Dispatching," *Signal Engineer* 3, no. 3 (1910): 3–4; C. A. Hammond, "Train Orders by Telephone," *Railroad Gazette*, December 30, 1892; Hay, "Beginnings of Telephone Train Dispatching," 55; Cooke, *Telegraphic Railways*, 2; Daniel C. McCallum, "Superintendents Report," March 25, 1856, in *Annual Report of the New York and Erie Railroad Company for 1855* (New York: New York and Erie Printing Office, 1856), 33–37, 39–41, 50–54, 57–59, quoted in Chandler, *Railroads*, 105.

47. S. Whinery, "Dispatching by Telephone in 1883," *Railway Age*, July 10, 1908.

48. G. K. Heyer, "Recent Progress in Telephone Train Handling," *Signal Engineer* 3, no. 6 (1910): 249.

49. Usselman, *Regulating Railroad Innovation*, 306.

50. "Railroad Telegraph Superintendents' Convention," *Railroad Gazette*, June 22, 1894; "The Telephone in Railroad Service," *Railroad Gazette*, July 28, 1899.

51. "Report of Committee on Telegraph," November 22, 1897, 1, 2, 7. Accession 1807/1810: Papers of the Association of the Transportation Officers of the Pennsylvania Railroad (henceforth PRR ATO Files), box 407, folder 8, Hagley Museum and Library.

52. "Report of Committee on Telegraph," November 22, 1897, 7, 8, 14. PRR ATO Files, box 407, folder 8, Hagley Museum and Library.

53. "Report of Committee on Telegraph," November 22, 1897, 16–17. PRR ATO Files, box 407, folder 8, Hagley Museum and Library.

54. Usselman, *Regulating Railroad Innovation*, 306; "Bell Telephone Loses," *New York Times*, February 28, 1901; D. H. Lovell to W. W. Atterbury. October 19, 1903. PRR ATO Files, box 415, file 263, folder 3—Movement of Trains by Telegraph, 2. See chaps. 3 and 4 of Beauchamp, *Invented by Law*, for an overview of the court cases related to various telephone patents in the United States.

55. G. F. J., "Telephone Train Dispatching," *Railroad Gazette*, January 4, 1901; Editorial, "Telephones in Train Dispatching and Other Railroad Work," *Railroad Gazette*, March 15, 1901. See Usselman, *Regulating Railroad Innovation*, chap. 8, for a discussion of Interstate Commerce Commission efforts to implement automatic block signaling requirements during the early twentieth century.

56. Editorial, "Railway Telephone Service," *Railway Age*, January 20, 1905; "Telephones on the B&O," *Railroad Gazette*, January 20, 1905; "Telephone Service on the Baltimore & Ohio," *Railway Age*, February 16, 1906.

57. Editorial, *Railroad Gazette*, July 19, 1901; Editorial, "Railway Telephone Service," *Railway Age*, January 20, 1905; Usselman, *Regulating Railroad Innovation*, 306.

58. Telegraph line builders typically used the earth to complete the electrical circuit, hence the term *ground return*. Telephone lines required a much higher quality electrical circuit for clear communication and thus builders had to employ fully "metallic" circuits.

59. "'Composite' Telephone Lines," *Railroad Gazette*, February 17, 1905.

60. Since railroads were setting up their telephone networks as private branch exchanges that did not interconnect with the switched public telephone network operated by the Bell System and various independent telephone companies, telephone officials such as AT&T president Theodore Vail seem not to have been particularly concerned about the telephone equipment utilized by the railroads. This is in marked contrast to Vail's efforts to tightly control telephone equipment used within the Bell System as part of his vision of "universal service." See John, *Network Nation*, chap. 10.

61. "The Cummings-Wray System of Handling Trains by Telephone," *Signal Engineer* 1, no. 8 (1909): 322–323; Editorial, "Growth of Telephone Train Dispatching," *Signal Engineer* 3, no. 3 (1910): 3–4.

62. W. W. Ryder, "Dispatching Trains by Telephone," *Signal Engineer* 1, no. 2 (1908): 55–56.

63. Usselman, *Regulating Railroad Innovation*, 307.

64. Editorial, "The Passing of the Railroad Telegrapher," *Signal Engineer* 1, no. 2 (1908): 67.

65. "Telephone Train Dispatching for Busy Line," *Signal Engineer* 2, no. 8 (1910): 249.

66. "Growth of Telephone Train Dispatching," *Signal Engineer* 3, no. 1 (1910): 3–4; ICC, *Twenty-Third Annual Report*, 49. In June 1910, Congress passed the Mann-Elkins Act, which expanded the ICC's power to regulate rail rates and business practices. The act also gave the ICC regulatory authority over interstate telegraphy and telephony. The act had little impact on railroad telegraph and telephone dispatching networks, however, since they were almost exclusively for railroad use and not interstate commercial communication entities. Furthermore, an institutional legal history of the ICC noted that few cases involving telegraph or telephone rate regulations appeared before the ICC in the decades after the passage of the Mann-Elkins Act. Sharfman, *Interstate Commerce Commission*, 110–111. Also see Childs, *Texas Railroad Commission*, for more information on railroad regulation at the state level in the late nineteenth century.

67. McIsaac, *Order of Railroad Telegraphers*, 19, 65, 239, 243–245.

68. Hay, "Beginnings of Telephone Train Dispatching," 59.

69. Ironically, in the 1950s, the American railroad industry was an early adopter of a new communication medium, microwave radio. In the case of the Southern Pacific Railroad, the firm set up the Southern Pacific Communications Company (SPCC) in 1970 to

oversee coast-to-coast expansion of the railroad's existing microwave network and eventually to lease excess capacity to commercial entities that wanted access to private, long-distance telephone networks for less money than AT&T was charging at the time. SPCC began marketing the service in 1973 but was forced to suspend operations after AT&T challenged the legality of private long-distance networks that interconnected to its local telephone exchanges. A coalition of private microwave providers led by MCI challenged the FCC decision and eventually won a court case in 1978 that permitted them to interconnect with the Bell System. SPCC resumed operations and began branding its private, switched telephone network as SPRINT (Southern Pacific Railroad Internal Network Telecommunications). For more information, see Strack, "SP and Sprint."

Conclusion

1. John, *Network Nation*, 164–170.
2. Thompson, *Wiring a Continent*, 443.
3. For more on the scholarly dangers of uncritically utilizing the concept of technology to describe and explain mechanical, institutional, and political-economic change over time, see Marx, "Technology."
4. H. D. Estabrook, "The First Train Order by Telegraph," *Baltimore and Ohio Employees Magazine* 1 (July 1913): 27–29, cited in Thompson, *Wiring a Continent*, 204.

Archival Sources

CALIFORNIA STATE RAILROAD MUSEUM, SACRAMENTO, CALIFORNIA

Railway & Locomotive Historical Society Collection

HAGLEY MUSEUM AND LIBRARY, WILMINGTON, DELAWARE

Pennsylvania Railroad Corporation, Lines East Corporate Records (including Lehigh Valley Railroad; New Jersey Railroad and Transportation Company; Camden & Amboy Railroad, Philadelphia, Wilmington & Baltimore Railroad, et al.), Accession 1807: Pennsylvania Railroad Corporation, Record Group 3, Soda House

Pennsylvania Railroad Corporation, Office of the Secretary, Accession 1807: Pennsylvania Railroad Corporation, Record Group 1, Soda House

Reading Railroad Records, Accession 1520: Reading Corporation, Soda House

LIBRARY OF CONGRESS, MANUSCRIPTS COLLECTION, WASHINGTON, DC

Andrew Carnegie Papers, 1835–1919

Herman Haupt Papers in the Lewis Muhlenberg Haupt Collection, 1861–1923

Samuel Finley Breese Morse Papers, 1793–1919

NATIONAL ARCHIVES, WASHINGTON, DC

Records of the Office of the Director and General Manager, Military Railroads United States; Records of the Office of the Quartermaster General, Record Group 92

NATIONAL MUSEUM OF AMERICAN HISTORY, KENNETH E. BEHRING CENTER, WASHINGTON, DC

Western Union Telegraph Company Records, NMAH.AC.0205

NEWBERRY LIBRARY, SPECIAL COLLECTIONS DEPARTMENT, CHICAGO, ILLINOIS

CB&Q Corporate Records, Series 33.1870.5.1: Contract File, 1856–1881

Illinois Central Corporate Records, Series +3.1: Minute Books and Indexes

WESTERN RESERVE HISTORICAL SOCIETY, MANUSCRIPTS COLLECTION, CLEVELAND, OHIO

Jeptha Homer Wade Family Papers

Periodicals

American Railroad Journal, New York: Railroad Journal Publishing Company.

American Railway Times, Boston: Railway Times Publishing Company.

Engineering News-Record: A Journal of Civil, Mechanical, Mining, and Electrical Engineering, New York: McGraw-Hill Publishing Company.

Journal of the Telegraph, New York: Polhemus & Landis, Publishers.
Locomotive Engineering: A Practical Journal of Railway Motive Power and Rolling Stock, New York: Angus Sinclair, Publisher.
The New York Times.
Railroad Gazette, New York: Railroad Gazette Publishing Company.
Railroad History, Westford, MA: The Railway & Locomotive Historical Society, Inc.
Railway Age, Chicago: The Railway Age Publishing Company.
The Scientific American, New York: Scientific American Publishing Company.
The Signal Engineer, Chicago: The Signal Engineer Company.
Telegrapher: Journal of Electrical Progress, New York: National Telegraph Union.
The Washington Post.

Primary Sources

Adams, Charles Francis, Jr. *Notes on Railroad Accidents*. New York: G. P. Putnam's Sons, 1879.

American Railway Association. *Proceedings of the General Time Convention and Its Successor the American Railway Association from Its Organization April 14, 1886 to October 11, 1893 Inclusive*. New York: The American Railway Association, 1893.

Anderson, J. A. *The Train Wire: A Discussion of the Science of Train Dispatching*. New York: The Railroad Gazette Publishing Company, 1891.

Board of Railroad Commissioners. *Third Annual Report of the Board of Railroad Commissioners*. Boston, MA: Wright & Potter, State Printers, 1872.

Butler, Benjamin F. *Private and Official Correspondence of Gen. Benjamin F. Butler, during the Period of the Civil War*. Vol. 3. Springfield, MA: Plimpton, 1917.

Carnegie, Andrew. *The Autobiography of Andrew Carnegie*. With John Charles Van Dyke. New York: Houghton Mifflin, 1920.

"Centennial of Train Dispatching Celebrated at Point of Origins and on Coast-to-Coast Telegraph Wire Hookup." *Train Dispatcher* 33, no. 11 (1951): 518–525.

Cooke, William Fothergill. *Telegraphic Railways; or, The Single Way Recommended by Safety, Economy, and Efficiency, under the Safeguard and Control of the Electric Telegraph: With Particular Reference to Railway Communication with Scotland, and to Irish Railways*. London: Simpkin, Marshall & Co., 1842.

Cooke, William Fothergill. *Treatise on Telegraph Law*. New York: Wm. Siegrist, 1920.

Eastern Railroad Association. Executive Committee. *Records of the New England Association of Railway Superintendents. Organized in Boston, Massachusetts, April 5, 1848. Dissolved, October 1, 1857*. Washington, DC: Press of Gibson Brothers, 1910.

Eighth Annual Report on the Statistics of Railways in the United States for the Year Ending June 30, 1895. Washington, DC: Government Printing Office, 1896.

Estes, George. Federal Writers' Project Interview with George Estes, November 28, 1938, 5–6, American Life Histories: Manuscripts from the Federal Writers Project, 1936–1940, Library of Congress, http://lcweb2.loc.gov/cgi-bin/ampage?collId=wpa2&fileName=29/2907/29070630/wpa229070630.db&recNum=19&itemLink=D?wpa:9:./temp/~ammem_jiTf::%2329070630002 0&linkText=1 (accessed August 1, 2018).

George, Charles B. *Forty Years on the Rail: Reminiscences of a Veteran Conductor*. Chicago: R. R. Donnelley & Sons, Publishers, 1887.

Gerstner, Franz Anton Ritter von. *Early American Railroads: Franz Anton Ritter von Gerstner's Die Innern Communicationen (1842–1843)*, ed. Frederick C. Gamst, trans. David J. Diephouse and John C. Decker. Stanford, CA: Stanford University Press, 1997.

Haupt, Herman. *Reminiscences of General Herman Haupt*. New York: Arno, 1981.

ICC (Interstate Commerce Commission). *Thirteenth Annual Report on the Statistics of Railways in the United States for the Year Ending June 30, 1900*. Washington, DC: Government Printing Office, 1900.

ICC (Interstate Commerce Commission). *Eighteenth Annual Report of the Interstate Commerce Commission*. Washington, DC: Government Printing Office, 1904.

ICC (Interstate Commerce Commission). *Twenty-Second Annual Report of the Interstate Commerce Commission*. Washington, DC: Government Printing Office, 1909.

ICC (Interstate Commerce Commission). *Twenty-Third Annual Report of the Interstate Commerce Commission*. Washington, DC: Government Printing Office, 1910.

ICC (Interstate Commerce Commission). *Twenty-Third Annual Report on the Statistics of Railways in the United States for the Year Ending June 30, 1910*. Washington, DC: Government Printing Office, 1912.

ICC (Interstate Commerce Commission). *Twenty-Fourth Annual Report of the Interstate Commerce Commission*. Washington, DC: Government Printing Office, 1911.

ICC (Interstate Commerce Commission). *Twenty-Fifth Annual Report of the Interstate Commerce Commission*. Washington, DC: Government Printing Office, 1912.

ICC (Interstate Commerce Commission). *Twenty-Ninth Annual Report on the Statistics of Railways in the United States for the Year Ending June 30, 1916*. Washington, DC: Government Printing Office, 1918.

Instructions for the Running of Trains, Etc. on the New York and Erie Railroad. New York: Press of the New York and Erie Railroad Company, 1857.

Johnston, William J. *Telegraphic Tales and Telegraphic History*. New York: W. J. Johnston, 1880.

Michigan Central Railroad Company. *Report of the Directors of the Michigan Central Railroad Company to the Stockholders* [for June 1856]. Boston: J. H. Eastburn's Press, 1856.

New York Central Railroad Company. *Annual Report of the New York Central Railroad Company*. Albany, NY: Weed, Parsons and Company, Printers, 1855.

O'Brien, John Emmet. *Telegraphing in Battle: Reminiscences of the Civil War*. Scranton, PA: Raeder, 1910.

Plum, William R. *The Military Telegraph during the Civil War in the United States*. Chicago: Jansen, McClurg & Company Publishers, 1882; reprint, New York: Arno, 1974.

Railroad and Warehouse Commission of Minnesota. *Twenty-First Annual Report of the Railroad and Warehouse Commission of Minnesota to the Governor for the Year Ending November 30, 1905*. St. Paul, MN: The Pioneer Press, 1906.

Reid, James D. *The Telegraph in America: Its Founders, Promoters, and Noted Men*. New York: John Polhemus, 1886.

Roosevelt, Theodore. "Third Annual Message," December 6, 1904. *The American Presidency Project*, by John T. Woolley and Gerhard Peters. http://www.presidency.ucsb.edu/ws/?pid=29545#axzz1UBETIiA8 (accessed August 1, 2018).

Rules and Regulations for the Government of the Transportation Department of the Pennsylvania Railroad. Philadelphia: E. C. Markley & Sons, 1874. http://prr.railfan.net/documents/RuleBooks/RulesAndRegulationsOfThePRR-1874.pdf (accessed October 12, 2018).

Rules and Regulations for the Government of the Transportation Department of the Pennsylvania Railroad. Philadelphia: E. C. Markley & Sons, 1882. https://babel.hathitrust.org/cgi/pt?id=hvd.hb4cp1;view=1up;seq=7 (accessed October 12, 2018).

Stickney, William, ed. *Autobiography of Amos Kendall*. New York: Peter Smith, 1949.

Talbott, E. H., and H. R. Hobart, eds. *The Biographical Dictionary of the Railway Officials of America*. Chicago: The Railway Age Publishing Company, 1885.

Taylor, Joseph. *A Fast Life on the Modern Highway: Being a Glance into the Railroad World*. New York: Harper & Brothers Publishers, 1874.

U.S. Congressional Serial Set, vol. 464. Washington, DC: Government Printing Office, n.d.

War of the Rebellion: Official Records of the Union and Confederate Armies. Washington, DC: Government Printing Office, 1880–1901.

Williams, W. *Appleton's Railroad and Steamboat Companion*. New York: D. Appleton, 1847.

Secondary Works

Aldrich, Mark. *Death Rode the Rails: American Railroad Accidents and Safety, 1828–1965*. Baltimore, MD: Johns Hopkins University Press, 2006.

Angevine, Robert G. *The Railroad and the State: War, Politics, and Technology in Nineteenth-Century America*. Stanford, CA: Stanford University Press, 2004.

Army, Thomas F., Jr. *Engineering Victory: How Technology Won the Civil War*. Baltimore, MD: Johns Hopkins University Press, 2016.

Bartky, Ian R. "Running on Time." *Railroad History* 159 (Autumn 1988): 18–38.

Bartky, Ian R. *Selling the True Time: Nineteenth-Century Timekeeping in America*. Stanford, CA: Stanford University Press, 2000.

Bartky, Ian R., and Carlene Stephens. "Comment and Response on 'The Most Reliable Time.'" *Technology and Culture* 32, no. 1 (1991): 183–186.

Beauchamp, Christopher. *Invented by Law: Alexander Graham Bell and the Patent That Changed America*. Cambridge, MA: Harvard University Press, 2015.

Bijker, Wiebe E., Thomas P. Hughes, and Trevor J. Pinch, eds. *The Social Construction of Technological Systems*. Cambridge, MA: The MIT Press, 1989.

Black, Robert C., III. *Railroad of the Confederacy*. Chapel Hill: University of North Carolina Press, 1952.

Calvert, James B. "Notes on Pennsylvania Railroad Operation and Signaling." https://mysite.du.edu/~jcalvert/railway/prr/prrsig.htm (accessed July 22, 2018).

Campbell, Thomas C., Jr. "The Wabash—The Gould Downfall." *Western Pennsylvania History* 33, nos. 1–2 (1950): 21–42.

Carey, James W. *Communication as Culture: Essays on Media and Society*. New York, NY: Routledge, 1988.

Cassale, John. "A Monument to Charles Minot," Telegraph History. http://www.telegraph-history.org/charles-minot/ (accessed July 6, 2018).

Chandler, Alfred D., Jr., ed. *The Railroads: The Nation's First Big Business—Sources and Readings*. New York: Arno, 1981.

Chandler, Alfred D., Jr. *Scale and Scope: The Dynamics of Industrial Capitalism*. Cambridge, MA: The Belknap Press of Harvard University Press, 1994.

Chandler, Alfred D., Jr. *The Visible Hand: The Managerial Revolution in American Business*. Cambridge, MA: The Belknap Press of Harvard University Press, 1977.

Childs, William R. *The Texas Railroad Commission: Understanding Regulation in America to the Mid-Twentieth Century*. College Station: Texas A&M Press, 2005.
Churella, Albert J. *Building an Empire, 1846–1917*. Vol. 1 of *The Pennsylvania Railroad*. Philadelphia: University of Pennsylvania Press, 2012.
Clark, John E., Jr. *Railroads in the Civil War: The Impact of Management on Victory and Defeat*. Baton Rouge: Louisiana State University Press, 2001.
Coe, Lewis. *The Telegraph: A History of Morse's Invention and Its Predecessors in the United States*. Jefferson, NC: McFarland, 2003.
Dilts, James D. *The Great Road: The Building of the Baltimore and Ohio, the Nation's First Railroad, 1828–1853*. Stanford, CA: Stanford University Press, 1993.
Downey, Greg. "Virtual Webs, Physical Technologies, and Hidden Workers." *Technology and Culture* 42, no. 2 (2001): 209–235.
Du Boff, Richard B. "Business Demand and the Development of the Telegraph in the United States, 1844–1860." *Business History Review* 54, no. 4 (1980): 459–479.
Du Boff, Richard B. "The Telegraph and the Structure of Markets in the United States 1845–1890." *Research in Economic History* 8 (1983): 253–277.
Dunlavy, Colleen A. "Organizing Railroad Interests: The Creation of National Railroad Associations in the United States and Prussia." *Business and Economic History* 19, series 2 (1990): 133–142.
Ellet, Elizabeth F. "Anson Stager." *Encyclopedia of Biography of Illinois*. Chicago, IL: Century Publishing & Engraving Co., 1892–1901.
Ely, James W., Jr. *The Railroads and American Law*. Lawrence: University Press of Kansas, 2001.
Field, Alexander James. "The Magnetic Telegraph: Price and Quantity Data, and the New Management of Capital." *Journal of Economic History* 52, no. 2 (1992): 401–413.
Fields, Gary. *Territories of Profit: Communications, Capitalist Development, and the Innovative Enterprises of G. F. Swift and Dell Computer*. Stanford, CA: Stanford University Press, 2004.
Gabel, Christopher R. *Railroad Generalship: Foundations of Civil War Strategy*. Fort Leavenworth, KS: Combat Studies Institute, 1997.
Gabel, Christopher R. *Rails to Oblivion: The Decline of Confederate Railroads in the Civil War*. Fort Leavenworth, KS: Combat Studies Institute, 2002.
Gabler, Edwin. *The American Telegrapher: A Social History, 1860–1900*. New Brunswick, NJ: Rutgers University Press, 1988.
Grodinsky, Julius. *Jay Gould, His Business Career, 1867–1892*. Philadelphia: University of Pennsylvania Press, 1957.
Harlow, Alvin F. *Steelways of New England*. New York: Creative Age, 1946.
Hay, Warren H. "The Beginnings of Telephone Train Dispatching." *Railroad History* 130 (1974): 55–60.
Hochfelder, David Paul. *The Telegraph in America, 1832–1920*. Baltimore, MD: Johns Hopkins University Press, 2012.
Howe, Daniel Walker. *What Hath God Wrought: The Transformation of America, 1815–1848*. New York: Oxford University Press, 2007.
Hungerford, Edward. *Men of Erie: A Story of Human Effort*. New York: Random House, 1946.

Israel, Paul. *From Machine Shop to Industrial Laboratory: Telegraphy and the Changing Context of American Invention, 1830–1920*. Baltimore, MD: Johns Hopkins University Press, 1992.

Jepsen, Thomas C. *My Sisters Telegraphic: Women in the Telegraph Office, 1846–1950*. Athens: Ohio University Press, 2000.

John, Richard R. "American Historians and the Concept of the Communications Revolution." In *Information Acumen: The Understanding and Use of Knowledge in Modern Business*, ed. Lisa Bud-Frierman, 98–110. London: Routledge, 1994.

John, Richard R. *Network Nation: Inventing American Telecommunications*. Cambridge, MA: Harvard University Press, 2010.

John, Richard R. *Spreading the News: The American Postal System from Franklin to Morse*. Cambridge, MA: Harvard University Press, 1995.

Kamm, Samuel Richey. "The Civil War Career of Thomas A. Scott." PhD diss., University of Pennsylvania, 1940.

Kieve, Jeffrey. *The Electric Telegraph: A Social and Economic History*. Newton Abbot, UK: David & Charles, 1973.

Kirkland, Edward Chase. *Men, Cities, and Transportation: A Study of New England History, 1820–1900*. Vol. 1. New York: Russell & Russell, 1948.

Kitchenside, Geoffrey, and Alan Williams. *Two Centuries of Railway Signaling*. Sparkford, UK: Oxford, 1998.

Klein, Maury. *The Life and Legend of Jay Gould*. Baltimore, MD: Johns Hopkins University Press, 1997.

Lord, Francis A. *Lincoln's Railroad Man: Herman Haupt*. Cranbury, NJ: Associated University Presses, 1969.

Marx, Leo. *The Machine in the Garden: Technology and the Pastoral Ideal in America*. London: Oxford University Press, 1964.

Marx, Leo. "Technology: The Emergence of a Hazardous Concept." *Technology and Culture* 51, no. 3 (2010): 561–577.

Marx, Leo, and Merritt Roe Smith, eds. *Does Technology Drive History? The Dilemma of Technological Determinism*. Cambridge, MA: The MIT Press, 1994.

McCrossen, Alexis. *Marking Modern Times: A History of Clocks, Watches, and Other Timekeepers in American Life*. Chicago: University of Chicago Press, 2013.

McDonald, Charles W. *The Federal Railroad Safety Program: 100 Years of Safer Railroads*. Washington, DC: Government Printing Office, 1993.

McIsaac, Archibald M. *The Order of Railroad Telegraphers: A Study in Trade Unionism and Collective Bargaining*. Princeton, NJ: Princeton University Press, 1933.

Mott, Edward Harold. *Between the Ocean and the Lakes: The Story of Erie*. New York: John S. Collins, 1899.

Murray, John. *Stokers and Pokers; or, The London and North-Western Railway, the Electric Telegraph, and the Railway Clearing-House*. John Murray, 1849.

Nonnenmacher, Tomas W. "Law, Emerging Technology, and Market Structure: The Development of the Telegraph Industry: 1838–1868." PhD diss., University of Illinois at Urbana-Champaign, 1996.

O'Connell, Charles F. "The United States Army and the Origins of Modern Management, 1818–1860." PhD diss., The Ohio State University, 1982.

O'Malley, Michael. *Keeping Watch: A History of American Time*. Washington, DC: Smithsonian Institution Press, 1990.

Overton, Richard C. *Burlington Route: A History of the Burlington Lines*. New York: Alfred A. Knopf, 1965.

Perrow, Charles. *Normal Accidents: Living with High-Risk Technologies*. New York: Basic Books, 1984.

Pietruska, Jamie L. *Looking Forward: Prediction and Uncertainty in Modern America*. Chicago: University of Chicago Press, 2017.

"Railroads and the Telegraph, Part I." *P.S.—A Quarterly Journal of Postal History* 22 (June 1984).

Rosenthal, Caitlin. "Big Data in the Age of the Telegraph." *McKinsey Quarterly* (March 2013). http://www.mckinsey.com/business-functions/organization/our-insights/big-data-in-the-age-of-the-telegraph (accessed July 7, 2018).

Russell, Andrew L. "'Industrial Legislatures': Consensus Standardization in the Second and Third Industrial Revolutions." *Enterprise & Society* 4, no. 4 (2009): 661–674.

Russell, Andrew L. *Open Standards and the Digital Age: History, Ideology, and Networks*. Cambridge: Cambridge University Press, 2014.

Salsbury, Stephen. *The State, the Investor, and the Railroad: The Boston & Albany, 1825–1867*. Cambridge, MA: Harvard University Press, 1967.

Sharfman, Isaiah Leo. *The Interstate Commerce Commission: A Study in Administrative Law and Procedure*. New York: The Commonwealth Fund, 1931.

Shaw, Robert B. *A History of Railroad Accidents, Safety Precautions, and Operating Practices*. Binghamton, NY: Vail-Ballou, 1978.

Shaw, Robert B. "Railroad Accidents and Passenger Safety." *Railroad History* 184 (Spring 2001): 22–36.

Silverman, Kenneth. *Lightning Man: The Accursed Life of Samuel F. B. Morse*. New York: Da Capo Press, 2004.

Solomon, Brian. *Railroad Signaling*. Minneapolis, MN: Voyageur, 2010.

Stephens, Carlene. "Astronomy as Public Utility: The Bond Years at the Harvard College Observatory." *Journal of the History of Astronomy* 21 (1990): 21–36.

Stephens, Carlene. *Time and Navigation: The Untold Story of Getting from Here to There*. Washington, DC: Smithsonian Books, 2015.

Stilgoe, John R. *Metropolitan Corridor: Railroads and the American Scene*. New Haven, CT: Yale University Press, 1985.

Stilgoe, John R. "Sounders and Silence: Some Isolated Train-Order Stations in Fiction." *Railroad History* 157 (Autumn 1987): 45–54.

Stover, John F. "Southern Railroad Receivership in the 1870s." *Virginia Magazine of History and Biography* 63, no. 1 (1955): 40–52.

Strack, Don. "SP and Sprint," UtahRails.net. http://utahrails.net/sp/sprint.php (accessed July 6, 2018).

Taylor, George Rogers, and Irene D. Neu. *The American Railroad Network, 1861–1890*. Cambridge, MA: Harvard University Press, 1956.

Thompson, Robert Luther. *Wiring a Continent: The History of the Telegraph Industry in the United States, 1832–1866*. Princeton, NJ: Princeton University Press, 1947.

Turner, George Edgar. *Victory Rode the Rails: The Strategic Place of the Railroads in the Civil War*. Lincoln: University of Nebraska Press, 1992.

Turner, Gregg M., and Melancthon W. Jacobus. *Connecticut Railroads . . . An Illustrated History*, ed. Oliver Jensen. New Haven, CT: The Connecticut Historical Society, 1989.

Usselman, Steven Walter. *Regulating Railroad Innovation: Business, Technology, and Politics in America, 1840–1920*. Cambridge: Cambridge University Press, 2002.

Ward, James A. *J. Edgar Thomson: Master of the Pennsylvania*. Westport, CT: Greenwood, 1980.

Ward, James A. *That Man Haupt: A Biography of Herman Haupt*. Baton Rouge: Louisiana State University Press, 1973.

Weber, Thomas. *The Northern Railroads in the Civil War, 1861–1865*. New York: King's Crown, 1952.

White, Richard. *Railroaded: The Transcontinentals and the Making of Modern America*. New York: W. W. Norton, 2011.

Wiebe, Robert. *The Search for Order, 1877–1920*. New York: Hill and Wang, 1967.

Wilson, Mark R. *The Business of Civil War: Military Mobilization and the State, 1861–1865*. Baltimore, MD: Johns Hopkins University Press, 2006.

Wolff, Joshua. *Western Union and the Creation of the American Corporate Order, 1845–1893*. Cambridge: Cambridge University Press, 2013.

Yates, JoAnne. *Control through Communication: The Rise of System in American Management*. Baltimore, MD: Johns Hopkins University Press, 1989.

Page numbers in *italics* refer to illustrations.

accidents, 152n20, 152n30, 172n66, 173n84; calculating number of, 173n83; dispatching and, 111, 117, 134, 169n2; excessive working hours and, 129, 134; mergers and, 106; Revere collision, 85–88; state and federal regulation, xii–xiv, 50, 85–89, 115, 128–36, 148, 153n30; telegraph and, x–xii, 15–16, 25–26, 38–39, 53–54, 83–88; telegraphers and, xiv, 77–78, 98–103, 117, 122, 127. *See also* safety

Adams, Braman B., 118

Adams, Charles Frances, Jr.: block signaling, 88, 90, 102, 111; railroad safety efforts, 93, 99, 102, 173n84; telegraphy and Revere collision, 85–88

air brakes, 10, 173n84

Allegheny Portage Railroad, 40

Allegheny Valley Railroad, 65

Allen, William F., 112–14

American Railroad Journal, 47

American Railway Association. *See* General Time Convention

American Railway Times, 36

American system of dispatching: compared with British system, xii, 83, 98–101, 168n96; criticisms of, xii–xiii, 88, 97–102; and dispatching chain, 78, 81, 100, 117, 119; origins of, 77, 81–82; single-order compared to double-order, 109–10, 116–18, 174n24; train accidents and, 78, 81–83, 88, 90–91, 97–101, 122, 125–26; Western Union and, 90–97, 102, 147–48

American Telegraph Company (ATC), 59, 61, 91

American Telephone & Telegraph (AT&T), 178nn43–44, 179n60, 180n69

American Union Telegraph Company, 95–97, 148

Anderson, John A., 108–10, 116–18, 174n27

Andrew, John, 73–74, 167n79

Annapolis & Elk Ridge Railroad, 57, 59, 95

Aquia Creek, 68–69, 74

Association of Railroad Telegraph Superintendents, 138

Association of the Transportation Officers (ATO). *See under* Pennsylvania Railroad (PRR)

Atlantic & Ohio Telegraph Company, 32–33, 40–43, 53, 94

Atlantic & Pacific Telegraph Company (A&P), 95–96, 171n54, 171n57

Bain, Alexander, 34, 37

Baltimore, 8, 40, 164n6; and Civil War, 56–61, 95, 164nn9–10, 164nn22–23; telegraph lines, 6, 17–19, 21, 23, 25, 32, 36, 48, 152n10, 154n61, 155n67; train stations, 17–19, 56; tunnels, 164n6

Baltimore & Ohio Railroad (B&O), xv, 40, 104, 158n15, 164n9, 164n13; and Civil War, 56–60, 63, 164n22; and development of American railroad operational management practices, 8–9, 11; and Hours of Service Act (1907), 134; and Morse telegraph line along right-of-way, 6–7, 17–19, 147; and telegraphic train dispatching, 19, 87; and telephonic train dispatching, 138–40; and Western Telegraph Company, 48

Baltimore & Ohio Railroad Company v. Interstate Commerce Commission (1911), 134

bankruptcy, 37–38. *See also* receivership

Bell, Alexander Graham, 137

Bell Telephone Company, 137, 139, 178n44. *See also* American Telephone & Telegraph (AT&T)
Berliner, Emile, 139–40
block signals: and American railroad operational management reforms, 83, 88, 90, 102, 111, 169n21, 176n59; cost to operate, 90; in operation, 86, 88–90, 98–99, 105, 169n2, 170n29. *See also* Camden & Amboy Railroad (C&A); Philadelphia & Reading Railroad (Reading); *and see under* Great Britain
Blodgett, George, 138
Boston, 8, 36, 83–85; railroad network centered on, 9; telegraph lines, 24–26, 156n94; timekeeping and railroads, 27–29, 157n104, 157n108, 158n23
Boston, Revere Beach & Lynn Railroad, 137
Boston & Maine Railroad (B&M), 27–28, 36–37
Boston & Providence Railroad (B&P), 9–11, 50
brakemen, 129, 175n39; average time of employment, 176n5; dangers for, 25, 128; duties of, 10, 33
Britain. *See* Great Britain
British system of train dispatching, xii, 14–16, 31, 43, 49, 98–101, 137, 168n96, 170n32; compared with American, xii, 83, 98–101, 168n96. *See also* Great Britain
Brooks, David, 43
Brooks, John W., 4, 26, 52–53, 163n121
brotherhoods, 3, 125, 128–29, 130–31. *See also* Order of Railroad Telegraphers (ORT)
Brown, J. C., 125
Brunel, Isambard K., 14, 16
Bull Run, First Battle of (1861), 61–62, 165n31
Butler, Benjamin F.: Civil War service, 56–57, 59; and J. Gould, 95
Butler Amendment (1879), 95–97, 148, 156n83, 170n40

Calkins, William Henry, 115
Camden & Amboy Railroad (C&A): block signaling, 88–90, 105, 170n37, 174n20; and Magnetic Telegraph Company, 21–23, 89
Cameron, Simon, 55, 57–60, 62, 64–66, 164n9, 164n13, 165n35
Canfield, Thomas, 65
capital investment: British railroads and, ix, 12; and dispatching practices, 49, 97, 117, 118, 140; federal government and, 6; and railroad communication infrastructure, xii–xiii, 47, 83, 87–88, 90–91, 136; in United States, ix, 8, 30, 35, 103, 170n38
Capitol. *See* United States Capitol Building
Carnegie, Andrew: Civil War service, 55, 57–61, 63–66, 72, 164n7, 165n27; Pennsylvania Railroad (PRR) manager, 63, 76, 167n88; telegraph clerk, 31–33, 159n36
Cassatt, Alexander J., 149, 172n65
Caton, John D.: business dealings with railroads, 51–52, 77, 163n112; and founding of Western Union, 52; midwestern telegraph firms, 50–52
Central Pacific Railroad, 92–93, 96
Central Railroad of New Jersey, 76
Chandler, Alfred D., Jr., 160n66, 161n86
Chapters of Erie (Adams), 85
Charleston & Savannah Railroad, 116
charters. *See* corporate charters
Chicago, 51, 75, 107, 136, 139, 178n44
Chicago, Alton & St. Louis Railroad (Chicago & Alton), 52
Chicago, Burlington & Quincy Railroad (CB&Q), 75; telegraph lines along right-of-way, 51–52; telegraphic train dispatching, 77, 117, 133; telephonic train dispatching, 121, 136, 139–43
Chicago, Kalamazoo & Saginaw Railway, 137–38
Chicago & Eastern Illinois Railroad, 103, 114
Chicago and Mississippi Telegraph Company, 51
Chicago & North Western Railway, 139, 163n116
City Point & Petersburg Railroad, 74
Civil War (1861–65): Battle of Gettysburg (1863), 76; Butler in, 56–57, 59; Cameron in, 55, 57–60, 62, 64–66, 164n9, 164n13, 165n35; challenges for civilian railroads in, 75–78; First Battle of Bull Run (1861), 61–62, 165n31; Lincoln in, 56–57, 62–63, 66, 69; McClellan in, 63–64, 68–69; McDowell in, 62, 67–69; and Military

Telegraph Corps, xi, 55, 59, 61, 63, 65, 78–79, 165n45; Overland Campaign and Petersburg Siege (1864), 74; Peninsula Campaign (1862), 68–69; Pratt Street Riot (1861), 56; Stanton in, 66–68, 71–74, 79, 166n68; and telegraphic train dispatching, xi–xiii, 48, 52, 54–56, 61–62, 66–67, 69–73, 75–80, 167n88; and United States Military Railroad, xi–xii, 55, 59–62, 65–69, 71–74, 78–80, 160n66, 165n27; Western Virginia Campaign (1861), 64

Cleveland, Columbus, Cincinnati & Indianapolis Railroad, 114

Cleveland & Cincinnati Telegraph Company, 52

communication, 55, 57–58, 60–62, 79, 131, 151n1 (under *Introduction*); British railroads and, 12–16; business, xi, xiii, 41, 91–97; and railroad operations, xiii–xv, 4–7, 32–34, 44–47, 105–11, 121–22, 136–46; and railroad safety, ix, 2–3, 85, 94, 97–102, 114–19; and telegraph patents, 41–43. *See also* American system of dispatching; telegraph and telegraphy; telephone and telephony

competition, xi, 95, 97, 104–5, 147–48

conductors, xi, 9–11, 28–29, 33–34, 37–38, 44–45, 175n39

Congress. *See* United States Congress

Conklin, David H., 37, 159n32

consolidation. *See* mergers

contracts, 34, 40, 64; disputes over, 94–97, 172n65; exclusive between railroad and telegraph companies, xiii, 3, 53–54, 91, 96–97, 148; patent licensing, 19–21, 31, 34–38, 41–43, 91, 94, 139–40, 146; and railroad rights-of-way, 21–25, 48, 51–54, 92–96, 148, 152n10, 162n92, 172n65, 178n43; and ultra vires concerns, 20, 23–24, 95, 156n83, 162n92, 170n40 (*see also under* telegraph lines). *See also* rights-of-way; *and under* Western Union

Cooke, William F.: and British railroads, 4, 7, 13–15, 26; development of indicator telegraph, ix, 4, 6, 12–13, 35, 152n3, 153n41, 154n44; ideas regarding telegraphic railroad operational management, 13, 15–16, 19, 30, 43, 47, 49, 137, 169n2

cooperation. *See* interfirm cooperation

Cornell, Ezra: and Minot, 35–38; and Morse, 17–18; New York and Erie Telegraph Company and, 35–38, 159n39; Western Union and, 52

corporate charters: federal, 96–97; in Massachusetts, 9; in New York, 35, 171n54; in Pennsylvania, 23–24, 40, 171n54; state issued, x, 9, 20–21, 95–96; in Virginia, 61. *See also* state legislatures

Crofutt, George, 1–2, 151n2 (under *Introduction*); commissioned painting *American Progress*, 2

Curtain, Andrew, 62

Delaware, Lackawanna & Western Railroad, 140

Delaware and Hudson Canal Company, 116–17

dispatchers. *See* train dispatchers

District of Columbia. *See* Washington, DC

double-tracking, 84; and railroad safety, 10, 12–13, 47, 50, 89, 105, 111, 141–42; and railroad traffic, 12–13, 47, 83, 87, 105, 111, 141–42

Eastern Railroad: resistance to telegraphic train dispatching, 84–85; and Revere collision, 83–84

Edison, Thomas, 95, 159n36

Ellsworth, Henry L., 154n41

Emerson, Ralph Waldo, 1, 151n1 (under *Introduction*)

employee rule books. *See* railroad rule books

employees. *See the titles of individual train and telegraph jobs*

engineers, 162n87; locomotive, 9–11, 43–44, 47, 105, 128–29, 176n7; and railroad construction, 8, 35, 40, 44; and railroad management, 19, 25, 89, 108, 160n45, 168n96

England. *See* Great Britain

Erie Canal, 35

Esch, John, 129; Hours of Service Act (1907), 129–32

financial panics: of 1857, 47; of 1873, 83, 102; of 1893, 125; of 1907, 172n65

firemen, 129

Fisher, J. B., 142

flagmen. *See* brakemen
France, 4–5, 147
Fredericksburg Railroad, 67–68, 74

Gale, Leonard, 17
Garrett, John, 164n13
General Time Convention (GTC): and Southern Railway Time Convention (SRTC), 111–12; and Standard Code of Train Rules (Uniform Telegraph Orders and General Rules), xiii–xiv, 103–4, 110–11, 115–20, 122, 128; and Standard Time, xiii, 112–14, *113*
Georgia Central Railroad, 160n45
Gould, George, 172n65
Gould, Jay: business dealings with railroads and Western Union, 95–97, 102, 147–48, 171n57, 172nn63–65; and New York & Erie Railroad (Erie), 85
Gray, Elisha, 137
Great Britain: block signals, 89, 98–100, 169n2; railroads in, ix, 173n83; relationship between railroad officials and telegraph promoters in, ix–x, 4, 6–7, 12–16, 26–27, 30, 154n44; train dispatching in, xii, 14–16, 31, 43, 49, 98–101, 237, 168n96, 170n32
Great Western Railway (GWR), 14, 16
Green, Norvin, 96, 170n41

Haines, Henry Stevens, 116
Hammond, Charles A., 137
Hammond, Charles Darius, 116–18
Harris, Robert, 77
Hartford & New Haven Railroad, 24
Harvard Observatory, 28–29, 157n104, 157n108
Haskins, Charles, 159n36
Haupt, Herman: ideas regarding railroad operational management, 69–72, 79, 81; and McCallum, 68–69, 167n79; and Pennsylvania Railroad (PRR), 32–33, 40–41, 165n35; and United States Military Railroad (USMRR), 68–74, 165n35, 166n68, 167n79
Henry, Joseph, 22
highways: public, xi, 18, 20, 23–24, 26, 95, 152n10; and telegraph rights-of-way, xi, 18, 20–24, 26, 95, 152n10; turnpikes, 21–24
Hours of Service Act (1907): and broken tricks (split shifts), 135–36; consequences for American railroad industry, xiv–xv, 120–22, 132; legislative history, 129–32; railroad industry response to, 133–36; and railroad telephony, 141–46
House, Royal: telegraph patent of, 34–35, 37, 41–42, 140
Hughett, Marvin, 52, 163n116
Hungerfold, Edward, 159n34

Illinois and Mississippi Telegraph Company (I&M), 51–52
Illinois Central Railroad (IC): commercial telegraph lines along right-of-way, 51, 76; telegraphic train dispatching, xv, 51–52, 76–77, 168n93
Illinois Central Telegraph Company, 51–52
interfirm cooperation, xvi, 3, 104–6, 110–12, 115, 145
Interstate Commerce Act (1887), 128
Interstate Commerce Commission (ICC): enforcement of Hours of Service Act (1907), 133–34, 136, 141; and interstate communication, 179n66; and passage of the Hours of Service Act (1907), 129–32; railroad safety advocacy, 128–29, 143, 172n66; and railroad telegraphy, 176n3, 176n7

Jackson, Andrew, 20
Jacksonian Democrats, x, 6, 19, 35
John, Richard R., xv–xvi
Journal of the Telegraph, 93–94

Kendall, Amos: frustration with railroads, 21–23; and Magnetic Telegraph Company (MTC), 21–25, 155n67, 158n23; and Morse, 20–21, 34; and New York, Albany & Buffalo Telegraph Company (NYA&B) line, 36

La Follette, Robert, 130–31; Hours of Service Act (1907), 130–32
Lake Shore & Michigan Southern Railway, 53, 114
Lamoreaux, Naomi, 158n11
Latrobe, Benjamin H., 7

Latrobe, John H. B., 7
Lebanon Valley Railroad, 76
Lehigh Valley Railroad, 122
Lincoln, Abraham, 56–57, 62–63, 66, 69
Liverpool & Manchester Railway, 13
London, 4, 13–15, 89
London & Birmingham Railway (L&B), 4, 13–14
London & Blackwall Railway (L&BR), 15
Long Bridge (Potomac River), 60–61, 165n24
Lyford, O. S., 103, 114

Mackie, John F., 126
Magnetic Telegraph Company (MTC), 21–25, 89, 154n61, 158n23
managers. *See* railroad managers
Manassas Gap Railroad, 165n31
Manhattan. *See* New York City
Mann-Elkins Act (1910), 165n31
Marx, Leo, 151n1 (under *Introduction*)
Massachusetts Board of Railroad Commissioners, 85–88, 90
McCallum, Daniel C.: and Haupt, 68–70, 166n65, 167n79; ideas regarding telegraphic railroad operational management, 44–47, 54, 56, 67, 71, 73, 79–81, 137, 160n66, 161n78, 161n86; managerial shortcomings of, 43, 47, 67–68, 72, 160n66; and Military Telegraph Corps, 67, 69–74, 79–80; and New York & Erie Railroad (Erie), 43–47, 54, 162n87, 168n99; and United States Military Railroad (USMRR), 55–56, 67–74, 79–81, 167n83
McCargo, David, 59
McClellan, George B., 63–64, 68–69
McCrea, James, 114–16, 118
McDowell, Irvin, 62, 67–69
MCI Communications Corp., 180n69
McLane, Louis, 19
McNeill, William Gibbs, and operational reforms: for Baltimore & Ohio Railroad, 8–9; for Massachusetts Railroads, 9–10
mergers, 103, 106, 125; New York & Erie Railroad (Erie), 38–39; New York Central & Hudson River Railroad (NYCR), 48; Pennsylvania Railroad (PRR), 107–8; Western Railroad, 11; Western Union, 52, 91, 94, 178n43
Merrick, Samuel, 41
Mexican War (1846–48), 55
Michigan Central Railroad, 4, 26, 52–54, 75, 159n36, 163n121
microwave radio, 179–80n69
Military Telegraph Corps: cost of a telegram, 168n104; employees, 62, 64, 79, 165n45; origins, xi, 59, 61, 63–65, 165n45; and United States Military Railroad (USMRR), xi–xii, 55–56, 60, 65–67, 69, 71–74, 79–80; wartime expansion, 55, 78–79
Milwaukee & St. Paul Railroad, 124
Minot, Charles: and Cornell, 35–38; and New York & Erie Railroad (Erie), 35–39, 41, 43–44, 47, 53, 159n28, 159n32, 159n36; telegraphic train dispatching, 35, 37–39, 43, 47, 159n28, 159n32
Missouri-Kansas-Texas Railroad, 144
Missouri Pacific Railway: Standard Time booklet, *113*
Moorhead, James K., 40–41
Morley, Robert F., 65
Morse, Samuel F. B.: and electromechanical indicator telegraph for St. Germain Railroad, 4–6, *5*; and first telegraph line along Baltimore & Ohio Railroad right-of-way, 6–7, 17–19, 23, 32, 147; and Kendall, 20–21, 34; telegraph patent of, 20–22, 31, 34–38, 41–42, 51, 94, 140, 152n3, 152n11, 153n41; vision for telegraph, x, 6, 14, 16, 19–21
Morse code, 34, 37, 86, 145, 158n12
Mott, Edward, 38

National Railway Convention (NRC), 105–6
National Road, 48
National Telegraph Act (1866), 92–93, 95–97, 156n83
networks: block signal, 89–90; Chappe optical telegraph, 4, 6; expansion of, 9, 43, 120, 125, 147, 172n65; government controlled, x, 6, 19; military railroad, 55–56, 59, 62–63, 65–67, 70, 73, 78–80; military telegraph, 55–56, 59, 62–64, 66–67, 72, 74, 78–81; overcapacity, 75, 81; radio, 148, 179n69; railroad, xii–xiii, xv, 1, 29, 44, 88, 111; railroad telegraph, 3, 14, 16,

networks (*continued*)
38–39, 48, 91–94, 170n40, 171n41; standardization of, 102–9, 116–17; telegraph, xi–xiii, 1, 29, 51–52, 144–45, 147–48, 179n66; telegraph network map of 1853, *18*; telephone, 121, 140–41, 145, 148, 179n60, 179n66. *See also* microwave radio; telegraph and telegraphy; telephone and telephony
New England Association of Railway Superintendents, 28–29, 36, 112, 157n103
New Haven & Hartford Railroad, 134
New Orleans & Northeastern Railroad, 137
New York, Albany & Buffalo Telegraph Company (NYA&B), 36, 48–49, 52, 162n92
New York & Erie Railroad (Erie): and McCallum; 43–47, 54, 162n87, 168n99; and Minot, 35–39, 41, 43–44, 47, 53, 159n28, 159n32, 159n36; and New York and Erie Telegraph Company, 35–38, 159n39; telegraphic train dispatching, 35, 37–39, 43, 47, 159n28, 159n32
New York and Erie Telegraph Company, 35–38, 159n39
New York and Western Union Telegraph Company. *See* New York and Erie Telegraph Company
New York Central & Hudson River Railroad (NYCR), xv, 53, 87, 158n15; expansion of its network, 103–4; Hours of Service Act (1907) and, 134, 140–41; and New York, Albany & Buffalo Telegraph Company (NYA&B), 48–49, 162n92; and railroad operating standards, 114
New York City: railroads, 9, 23, 35, 37–39, 107; telegraph lines, 19, 21–26, 29, 36, 156n94, 158n23
New York Railroad Journal, 13
New York Times, 131
Nicolls, Gustavus A., 25, 49, 76
Norfolk & Western Railroad, 139
Northern Central Railway, 59, 75, 164n9
Notes on Railroad Accidents (Adams), 88

O'Brien, John Emmet, 59
Old Colony Railroad, 125–26
operating procedures. *See* railroad rule books
Orange & Alexandria Railroad (O&A), 60–62, 65, 67–68, 72, 165n31
Order of Railroad Telegraphers (ORT): formation of, 120–21, 124–25, 135; and Hours of Service Act, 131–32, 134, 136; member card for, *121*; and telephone, 142–45; and train dispatchers, 125–27
Order of Railway Telegraphers of North America. *See* Order of Railroad Telegraphers
organizational hierarchy: challenges to, xi, xiv, 34, 81, 101, 161n86; diagram by McCallum, *46*, 161n78; and railroad management, 8, 106, 108, 119–20, 122, 138, 147, 161n78, 161n86; spatial, 9, 11–12, 15, 77–78, 108–9, 125, 137. *See also* Cooke, William F.; McCallum, Daniel C.; railroad managers; railroad operational management in the United States; superintendents
O'Rielly, Henry: and Atlantic & Ohio Telegraph Company, 32, 40; ideas about telegraphic train operational management, 49; and Magnetic Telegraph Company, 21, 23–24, 155n67
Osborn, William, 77, 168n93

Pacific & Atlantic Telegraph Company (P&A), 94–95, 171n54
Pacific Railway Acts (1862 and 1864), 96–97
Parliament (Great Britain), 169n2
patents, xi, xv, 34, 139–40, 178n44. *See also* telegraph patents
Paterson & Hudson River Railroad, 39
Pennsylvania Railroad (PRR), 78; and Association of the Transportation Officers (ATO), 107–10, 114, 138–40, 174n17; and Atlantic & Ohio Telegraph Company (A&O), 32–33, 40–43, 53, 94; and Civil War, 55–59, 63, 65, 75–76, 87; conflict with Western Union, 92–97, 172n65; construction of, x, 39–40, 69; evaluation of Morse and House telegraph patents for railroad telegraph line, 41–43; and General Time Convention, 114–19; map of its rail lines in 1892, *107*; rule books of 1874 and 1882, 108–9, 174n20; standardization efforts, 103–4, 107–10; Telegraph Committee of, 41–43, 138–40; telegraphic train dispatching,

xi–xii, 31–34, 47, 54, 89–90, 174n24; and telephone, 121, 138–40, 142–43
Pensacola Telegraph Company v. Western Union Telegraph Company (1877), 156n83
Philadelphia, 19, 32, 56, 58; Camden & Amboy Railroad (C&A), 89–90, 174n20; Magnetic Telegraph Company (MTC), 21–23, 155n67; Pennsylvania Railroad (PRR), 39–40, 42–43, 107, 143; Philadelphia & Reading Railroad (Reading), 49, 76, 170n29
Philadelphia, Wilmington & Baltimore Railroad (PW&B): during Civil War, 56, 59, 75; and Magnetic Telegraph Company, 23–24
Philadelphia & Reading Railroad (Reading), 61; block signals, 170n29; and telegraph, 25, 49, 54, 76
Pitcairn, Robert, 116–18
Pittsburgh: and Atlantic & Ohio Telegraph Company (A&O), 23, 32, 40, 43; and Pennsylvania Railroad (PRR), 32–33, 40, 42–43, 63, 76, 107–8, 139, 172n65
Pittsburgh, Fort Wayne & Chicago Railway (PFW&C), 75
Pope, Franklin L., 90
Potomac Creek, 69
Potomac River, 59–62, 65, 68, 74, 164n23, 165n24
Potomac Run Bridge, 69
Providence & Worcester Railroad (P&W), 49–50

Railroad Gazette, 98–101, 106–7, 109–10, 120, 123–25, 132–34, 140, 168n96, 174n27
railroad managers, 34, 40–41, 43–44; and American system of dispatching, 81–83, 88, 90–91, 97–102, 122, 125; in Britain, ix, 6, 12–16, 30, 89; and Civil War, 55–63, 65–81, 87; conflicts with commercial telegraph firms, xiii, 26–27, 93; development of operational rules and regulations, 7, 9–12, 28–30, 46, 50, 70–71, 108–10, 114–19, 147, 160n66, 161n86; distrust of telegraphers and train dispatchers by, xiv–xv, 3, 122–27, 132–34; reluctance to adopt telegraphy for operational management, 1–3, 16–17, 19, 29–30, 39, 48–52, 54, 76–79, 83–88, 147; and standardization, 103–19; use of telegraphy for operational management, x, 31–39, 47–50, 52, 54–59, 65–67, 71–72, 75–83, 89–93, 97–103, 109–10, 115–19, 147–49, 161n86; use of telephony for operational management, xv, 121–22, 136–46, 148. *See also* American system of dispatching; General Time Convention (GTC); railroad operational management; railroad rule books; standards; superintendents
railroad operational management in the United States: Civil War era, xi–xii; 55–56, 59, 67–73, 75–80; development of, ix–x, 7–12, 27–30, 32–33, 35–54, 147–48, 152n20, 153n28; post–Civil War challenges, xii–xiii, 81, 83–88, 97–102; standardization efforts, xiii–xiv, 103–9; telephone and, xiv–xv, 121–22, 136–46, 148
railroad rule books, 40–41, 43, 70–71, 79, 81; accidents and, 15–16, 27, 38, 84–87, 98–101, 106, 120, 122–23, 127, 153n30; development of, 8–13, 103–5, 109–10, 114–18, 138–39, 153n25, 153n28; and telegraphy, 16, 38–39, 44–47, 52–54, 77, 81, 85–87, 108–9, 116–18, 168n99, 174n20, and timekeeping, 27, 29, 88, 147
railroad telegraphers. *See* telegraphers
railroads and telegraphy. *See* telegraphy and railroads
railroads and timekeeping. *See* timekeeping and railroads
Railway Age, 109–10, 115, 127, 130, 136–37, 140
Railway Association of America, 101
Ralph, Joseph E., 115
receivership, 103, 162n88, 170n39. *See also* bankruptcy
regulation, xiii, 64; federal, xiv, 3, 88, 99, 115, 128–34, 136–37, 141–46, 148, 170n40, 170n66; and railroad management, 40; railroad operating rules and, 8–9, 29, 105–6, 114, 120, 138, 153nn28–29; and safety, 11, 16, 38, 85–86, 123; state, 50, 87–88, 99, 148
Reid, James D., 22, 25, 36, 48, 52–53
Republican Party, 129, 131
Rice, Reuben N., 53
Richmond & Petersburg Railroad, 74

rights-of-way, 20; allegorical, 1; contested, x, 2, 7, 16–17, 20–27, 155n71; depiction of telegraph lines along railroad right-of-way, 2, *82*; public, 20–21, 23, 25–26; railroad, 10, 61, 144; and telegraph, ix, 6, 17–18, 31, 40–43, 48–54, 82, 144, 152n10, 156n83; and telephone, 137; and Western Union, xiii, 3, 90–97, 148, 151n3 (under *Preface*), 163n122, 172n65, 178n43. *See also* highways
Rockefeller, John D., 172n65
Rock Island Railroad, 124
Roosevelt, Theodore, 128–32
rule books. *See* railroad rule books
Rutland & Washington Railroad, 65
Ryder, W. W., 141–42

safety, 27, 62, 132, 137, 142, 147–48, 173nn83–84; debates over, 97–101; public, 25, 87–91, 93–94, 169n2; and railroad operations, ix–x, 3, 7–8, 10, 12, 19, 27–30, 83, 152n20, 153n30, 157n108, 170n29; and standards setting, 105–6, 109–11, 115–20; and telegraphers, 122–23, 125–29; telegraphy and, xii–xiii, 2–4, 13, 15–16, 38–39, 43, 45, 48–50, 53, 56, 75–76, 80–83; telephone and, 138–39, 143, 145. *See also* accidents
Safety Appliance Act (1893), 128
San Antonio & Aransas Pass Railroad, 139
Sanford, Edward S., 59, 61
Santa Fe Railroad, 134
Scott, Thomas A.: Civil War service, 55, 57–67, 72–73, 79, 164n13, 164n15, 165n27, 165n35, 165n49; and Pennsylvania Railroad (PRR), 32–33, 55, 57–58, 63, 157n4, 165n35
Selden, Charles L., 138
Shaw, Robert B., 9
Signal Engineer, 142–43
signaling, 85, 118, 123, 142, 145, 170n29; automatic, 111, 140, 144, 176n59, 179n55; electrical, xiii, 4–5, 108, 119, 144; telegraphic, 15, 83, 86, 88–90, 98–101, 105, 119, 169n2, 169n21, 174n20; train crew operated, 10, 33, 114, 175n39; uniform, 103, 105–106, 108, 114–16, 120; voice operated, 13
signalmen, 89–90, 105, 169n2
Smith, Francis O. J.: frustration with railroads, 25–26; and Morse, 17, 20, 154n56; and Morse telegraph patent, 34, 36–37, 158n23; telegraph line construction in New England, 24–26, 29, 156n94
Smith, William H., 11, 27, 153n28
Southern Pacific Railroad (SP), 121, 134; and creation of Southern Pacific Railroad Internal Network Telecommunications (SPRINT), 179–80n69
Southern Railroad, 130, 133, 139
Stager, Anson: and exclusive Western Union railroad right-of-way contracts, 53, 163n122; and Military Telegraph Corps, 63–65, 67, 72–74, 79, 165n45
Standard Code of Train Rules. *See under* General Time Convention
standards, 9, 11–12, 104–5, 128, 161n86, 173n2; timekeeping, 27–29, 50, 112–13, 147, 156n97, 175n34; train crew signals, 114–15; train dispatching and telegraphy, xiii–xiv, 3, 34–36, 39, 43–45, 103–4, 106–11, 115–20, 122, 139, 145. *See also* General Time Convention; National Railway Convention; Pennsylvania Railroad (PRR): standardization efforts; timekeeping and railroads
Stanton, Edwin M., 66–68, 71–74, 79, 166n68
state legislatures, x, 21, 23; Connecticut, 25; Massachusetts, 25, 73–74, 153n30; New Jersey, 23; Pennsylvania, 40; Rhode Island, 50
station agents: xiii, 11–12, 27–29, 84, 106, 123; and telegraphy, xi, 4, 14–15, 37–38, 42, 72, 86, 114, 159n28; and telephony, 138–39, 141
Stearns, John O., 76
Stephenson, Robert, 14
St. Louis, 51–52, 107
St. Louis, Iron Mountain & Southern Railway, 101, 125
Stockton, Robert, 22
Stone, Henry B., 117
Stonington Line, 29, 83
street railroads, 164n6, 164n23
strikes, 44, 47, 124–25
Strouse, David, 59–61, 63–64, 66
superintendents, 19, 35–38, 63, 126, 130, 165n27, 166n65; railroad operations, 9–12, 15, 31–33, 40–47, 53, 67–69, 76–78, 87, 89,

98, 122–23; and standardization, 28–29, 103, 108–10, 112, 114–16, 118, 157n103; telegraphy, 31–33, 41–47, 53, 76–78, 87, 98, 122–23; telephony, 137–38, 141–42. *See also* General Time Convention (GTC); New England Association of Railway Superintendents; railroad managers
Supreme Court. *See* United States Supreme Court
systemization, 9, 12, 16, 33, 44–47, 85–88, 93, 101, 107–10, 114–19, 148. *See also* Adams, Charles Frances, Jr.; American system of dispatching; Cooke, William F.; McCallum, Daniel C.; Welch, Ashbel

technology, 149, 151n1 (under *Introduction*), 180n3
telegraph and telegraphy: Chappe system, 4, 6; during Civil War, 55, 59–67, 72–74, 78–80; in Europe, ix, 4–6, 12–16, 151n1 (under *Preface*), 154n44, 169n2; indicator, 4, 7, 13–14, 35, 89, 152n3; nationalization of, 147, 151n1 (under *Preface*); patent disputes, xi, 34–35, 37, 41–43, 95, 146, 153n41; quadruplex, 95; symbolism of, 1–2, 148–49; telegraphone, 141; United States government and, x, xiv, 6–7, 19, 21, 95–97, 156n83, 170n40, 179n66. *See also* telegraphers; telegraph lines; telegraph office; telegraph patents
telegraphers: female, 176n5; labor relations, xiv, 120–22, 124–25, 127, 131–36, 143–46, 148; safety concerns, xiv, 3, 38, 100–101, 117–20, 122–27, 130–32; train dispatching, 32–34, 38, 44–45, 52, 59, 115–16, 122–28, 134–36, 144–45; wages, xiv, 123, 126, 133–34, 176n7; working conditions, xiv, 62, 120, 122–23, 135–36, 165n45
telegraph lines: allegorical, 1; in Britain, 13–16, 151n1 (under *Preface*), 154n44; construction of, x, 6–7, 20–27, 36–37, 51–59, 63–65, 91–92, 154n57, 155n67, 171n46, 179n58; construction cost, 168n1; depiction of, along railroad right-of-way, *2*, *82*; hazards for railroads, x, 2, 22, 24–25, 27, 156n94; and railroad rights-of-way, ix–xi, 4, 6–7, 13–19, 21–27, 40–43, 48–49, 52–54, 91–97, 147–48, 151n3 (under *Preface*), 156n83, 163n122; state and federal laws regarding, 20–21, 23, 25–26, 92, 95–97, 156n83, 170n40; ultra vires concerns, x, 2–3, 20, 23–24, 95, 162n92, 170n40
telegraph office: commercial, 25, 29, 32; government, 61–62, 64; railroad, xiv, 32–33, 44, 54, 76, 130, 133, 135, 141
telegraph operators. *See* telegraphers
telegraph patents: of Bain, 34, 37; compared, 34–35, 41–43, 140, 152n3, 153–54n41; of Cooke and Wheatstone, 4, 6–7, 12–16, 35, 152n3, 153n41, 154n44; Ellsworth and, 154n41; of House, 34–35, 37, 41–42, 140; of Morse, x–xi, 4, 6, 21, 31, 36–38, 51
telegraph school, 134–35, 146; advertisement for, *135*
Telegrapher, The, 81, 84–85, 92–93, 100–101
telephone and telephony, 179n58, 179n66; American Telephone & Telegraph (AT&T), 179n60, 180n69; Hours of Service Act, xiv, 3, 121–22, 137, 141–43; and railroads, xiv, 3, 121–22, 136–46, 148; Western Electric, 178n44, 179n60
Texas & Pacific Railway, 157n4
Thomas, Edward B., 114
Thompson, Robert L., 19
Thomson, Frank, 108
Thomson, J. Edgar, 32, 40–41, 57–58, 108, 160n45, 161n86, 164n9, 164n13
Tillotsin, Luther, 38
timekeeping and railroads: civil time, 28, 112, 157n104, 175n38; railroad time standards, xiii, 27–29, 36, 49–50, 112–15, 147, 156n97, 157n104, 175n34, 175n38; telegraphy and, 28–29, 38–41, 44, 70–72, 77, 79, 111, 157n108, 175n34; timetables, 9–12, 15–16, 31–32, 47, 50, 52–54, 84, 87, 101, 153n28; watches and clocks, 28–29, 49–50, 70–71, 112, 147, 157n104, 157n108
timetables. *See under* timekeeping and railroads
traffic: 35, 44, 59–61, 110; block signaling, xii, 88–90, 111; on British railroads compared to American railroads, ix, 12, 30, 89, 147; increase during Civil War, xii, 63, 75–78, 81, 87; and railroad telegraph, xii–xiii, 33, 39, 49, 54, 76–78, 81, 101, 111, 120, 122, 130, 134; and railroad telephone, 122, 138,

traffic (*continued*)
140–41, 144–45; safety concerns, 3, 8, 10, 16, 33, 38, 50, 70, 83–85, 101–2, 117, 120, 134; and telegraph messages, 6, 18, 36, 41, 53, 72, 91, 97, 104, 148; volume of, 28, 31, 45, 49, 54, 103–4, 130, 138, 147
train dispatchers, 167n88, 176n3; operational dangers due to, xii, 3, 77–78, 81, 97–101, 109, 111, 115–16, 119–20, 122–23, 125–27, 134, 136; and railroad operations, 4, 31, 52, 70–72, 76–77, 129; and telephone, xiv, 122, 138–39, 141–43; and train orders, 93–94, 98–99 163n3, 174n24; wages, xiv, 126–27, 130, 143, 176n7; working conditions, xii, xiv, 71–73, 81, 111, 121, 125–27, 129–32. *See also* American system of dispatching; Order of Railroad Telegraphers (ORT): train dispatchers; Train Dispatchers' Association
Train Dispatchers' Association, 126–27
train orders. *See* American system of dispatching
Train Wire, The (Anderson), 110, 118
transcontinental railroads, 151n2 (under *Introduction*), 169n3
Tubbs, Frederick, 77

ultra vires. *See under* contracts; telegraph lines
Union Pacific Railroad, 92, 96, 134, 169n3, 172n63
unions. *See* brotherhoods
United States Army, x, 95, 170n40; and Civil War, 57, 60, 63, 68, 73–74, 79; Engineer Department of, 8. *See also* United States Military Railroad (USMRR); United States Military Telegraph Corps; United States War Department
United States Capitol Building, 17–18, 58, 130
United States Congress, xiv, 112, 128, 173n84; and Civil War, 60, 62, 66–67, 74, 105, 167n83; and telegraph, x, 6–7, 19, 21, 95, 97, 129–33, 141, 170n40, 179n66. *See also* Butler Amendment (1879); Hours of Service Act (1907); Interstate Commerce Act (1887)
United States Military Railroad (USMRR): 55, 73, 78–80, 160n66; Construction Corps, 69, 74; creation of, xi–xii, 59–60, 65–67, 165n27; and Military Telegraph Corps, xi–xii, 55–56, 60, 65–67, 69, 71–74, 79–80; operations in northern Virginia, 60–61, 65, 67–69, 71–72; operations in the Department of the Cumberland, 71, 73–74; operations in the Department of the Rappahannock, 68–69; reorganization of, 65–66, 73–74; support for US Army operations, 62, 67–69, 74
United States Military Telegraph Corps. *See* Military Telegraph Corps
United States Post Office Department, x, 19, 20, 22, 154n61
United States Supreme Court, 34, 96, 134, 156n83
United States Telegraph Company, 91
United States War Department, 35, 57, 59–68, 70, 73, 105; and Quartermaster Department, 57, 64–67, 73–74, 165n45; and southern railroads, 60, 67. *See also* Civil War; United States Army; United States Military Railroad (USMRR); United States Military Telegraph Corps
Usselman, Steven W., xv, 158n11, 176n59

Vail, Alfred, 17–18
Vail, Theodore, 179n60
Van Rensselaer, R. S., 89
Vermont and Boston Telegraph Company, 157n108
Virginia Central Railroad, 61

Wabash Railroad, 96, 115, 172n65
Wade, Jeptha H., 26, 52–53, 163nn121–22
Wade, Kirtland H., 114–16
Washington, DC, 132, 165n24; and Civil War, 56–62, 64, 67–69, 95, 164n15, 164n23; telegraph lines, 6, 17–19, 21, 23, 32, 36, 152n10
Welch, Ashbel: block signals, 88–89, 170n37, 174n20; standardization of railroad operating practices, 105–6
Wells, W. W., 101
Western New York & Pennsylvania Railway, 137
Western Railroad of Massachusetts, 11–12, 25, 27–28, 153n30

Western Telegraph Company, 48
Western Union Telegraph Company (WU), 1, 63–64, 121, 149, 159n36; advertisement for, *144*; and American Telephone & Telegraph (AT&T), 178n43; and exclusive railroad right-of-way contracts, xi, 3, 52–54, 91–93, 120, 147–48, 151n3 (under *Preface*), 163n122, 171n46; J. Gould and, 95–97, 102, 147–48, 171n57, 172n63, 172n65; and Pennsylvania Railroad (PRR), 94–97, 172n65; and railroad safety, 93–94, 97, 102, 148; railroad telegraph service, xiii, 52–54, 82–83, 90–94, 163n121, 170n41
Wheatstone, Charles: and British railroads, ix, 4, 7, 12–16, 26; joint telegraph development with Cooke, 4, 6, 12–13, 35, 152n3, 153n41, 154n44
Whig Party, 6, 19
Whinery, Samuel, 137
Whistler, George W., 9–12, 27
White, Richard, 2, 157n4, 161n86, 169n3
Willard, Simon, Jr., 157n104
Wright, Phineas P., 114

Yates, JoAnne, xv, 168n93